The Living World of the Sea

The WORLD of SURVIVAL

The LIVING WORLD of the sea

Bernard Stonehouse

HAMLYN
London · New York · Sydney · Toronto

Acknowledgements

Photographs

Ardea, London 19, 74; Ardea – J. A. Bailey 22 bottom; Ardea – Liz and Tony Bomford 63 bottom; Ardea – John Clegg 75 top; Ardea – Kenneth Fink end papers; Ardea – Ken Hoy 47; Ardea – Jane I. Mackinnon 61 top right; Ardea – John Mason 18; Ardea – P. Morris 42 bottom, 59 top left; Ardea – B. L. Sage 46 inset; Ardea – Ron and Val Taylor 58, 77 bottom, 91; Ardea – Valerie Taylor 54, 56 top, 59 top right, 65 bottom, 77 top; Ardea – Richard Vaughan 11; Biofotos, Farnham 12-13, 13 top, 14, 19 inset, 20, 22 top, 23, 24, 25, 26 top, 27, 28 top, 28 bottom, 29 top, 29 bottom, 30, 31 top, 32, 33 top, 34, 35 top, 35 bottom, 36 top, 37 bottom, 40 top, 41 top, 41 bottom, 42 top, 43 top, 45 top, 45 bottom, 46, 47 inset, 51 top left, 51 top right, 51 bottom, 55 top, 55 bottom, 56 bottom, 69 bottom, 90 top; Biofotos – Ian Took 57, 59, 61, 65; Biofotos – S. Summerhays 57, 89; Bruce Coleman – Jen and Des Bartlett 80-81; Bruce Coleman – Jane Burton 31, 53, 76; Bruce Coleman – Neville Coleman 63; Bruce Coleman – Eric Crichton 26; Bruce Coleman – D. Houston 81; Bruce Coleman – Jon Kenfield 48; Robert Harding Associates, London 10, 11 inset; Robert Harding Associates – Jeffrey Craig 38-39; Robert Harding Associates – Walter Rawlings 11; N.A.S.A., Washington, D.C. 6; Oxford Scientific Films 17 top, 21, 33 bottom, 43 bottom, 44, 50, 62 top, 67, 68, 69 top, 70 top, 70 bottom, 71, 72-73, 73 bottom left, 73 bottom right, 85 top left, 85 top right, 85 bottom, 86, 87 bottom, 88, 89 bottom; Oxford Scientific Films/Animals Animals – Zia Leszczynski 52 top, 52 bottom, 64-65; Oxford Scientific Films – J. A. L. Cooke 48 top; Survival Anglia Ltd 9 top; Survival Anglia Ltd – Jen and Des Bartlett front jacket, 66, 75 bottom left, 78, 92-93, 93; Survival Anglia Ltd – Les Bartlett for Jen and Des Bartlett 79; Survival Anglia Ltd – Moira Borland title page, 92; Survival Anglia Ltd – Rod and Moira Borland 13, 60, 79; Survival Anglia Ltd – Eva Cropp back jacket; Survival Anglia Ltd – Jeff Foott 9 centre, 17 bottom, 40 bottom, 61 top left, 90 bottom, 94 top, 94 bottom; Z.E.F.A. – G. Marche 16; Z.E.F.A. – K. Weck 38-39; Z.E.F.A. – Warren Williams 49; Iceland Tourist Board 83; N.H.P.A. – James Tallon 76 top; Seaphot – Bill McFarland 7 centre; Seaphot – Peter David 87 top; Syndication International 94.

Illustrations

Creative Cartography Ltd.; David Eaton; Roger Full; Mike Woodhatch.

Published 1979 by
The Hamlyn Publishing Group Limited
London · New York · Sydney · Toronto
Astronaut House, Feltham, Middlesex, England
Text © Copyright Bernard Stonehouse 1979
Illustrations © Copyright The Hamlyn Publishing Group Limited 1979
WORLD OF SURVIVAL
Registered Trademark

ISBN 0 600 35560 8

Printed in Italy

Contents

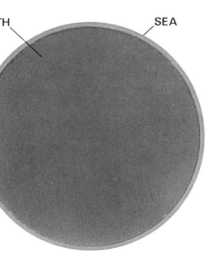

A world of water

Human beings are creatures of the land. We tend to think of our planet as a world of solid ground, a green world, covered with trees, shrubs and grass. But when astronauts glance back at the Earth from outer space, the planet seems mainly blue. Part of this is the blue of atmosphere, seen against the Earth's reflecting surface; most of it is the blue of the oceans. Oceans cover seven-tenths of the Earth's surface – almost three-quarters. We sometimes forget just how little of the world we can actually walk on.

Similarly, when we think of animals, we tend to think first of land animals: tigers and elephants, beetles and cows, earthworms, snails and sparrows. These are very interesting animals, and the ones with which we are most familiar. However, to someone looking in from outer space, they are merely the special kinds of animals that live crowded together with Man on the dry three-tenths of the Earth's surface. What

about the majority – the millions upon millions of animals, of all shapes and sizes, that live in the sea?

This book is about the enormous numbers of marine animals, just as interesting, just as lively and spectacular in their own ways, as animals of the land. It is mainly about the animals of the oceans, but it deals also with the plant life that supports them, and with the whole wet, salty world that is their home.

Looked at from outer space, the oceans give little impression of their true size. They cover a large area, but are spread thinly. If we scaled the world down to a model two metres (two yards) in diameter, we could represent the sea with a film of water averaging only one millimetre (about one twenty-fifth of an inch) thick. This is still a vast amount of ocean. The total volume of water in the seas is 15 times greater than the volume of land above sea

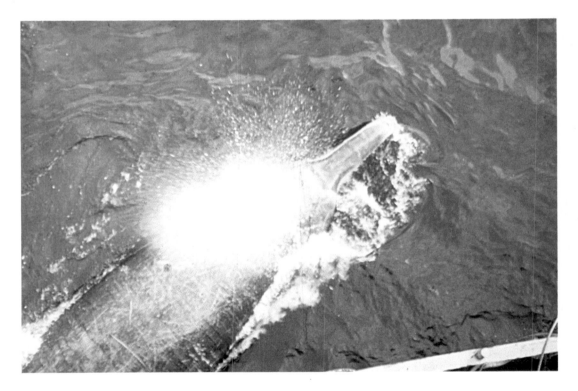

level. If we bulldozed all the land into the sea, and levelled it off under water, we should end up with a single ocean covering the whole of the Earth's surface, and averaging more than three kilometres (two miles) deep. As they stand at present, the depths of the oceans average five times the mean height of the land, and the deepest ocean trenches are far deeper than Mount Everest is high.

All this enormous volume of water contains living creatures. Some parts of it, like the deserts on land, have only a few plants or animals over a vast distance. Other parts of the ocean – some areas of the seabed, for example – are so rich in organisms that it would be difficult to find space between them.

Marine plants and animals come in all sizes. The smallest among them – bacteria, for example, and the individual plant cells that colour surface waters – are only a hair's-breadth across. However, these cells make up in numbers what they lack in size. Tiny as they are, there may well be millions of them in each small area of water, and they play very important roles in the lives of other sea creatures. If the surface-living plant cells died out, the rest of the ocean would die with them. At the other end of the scale are the biggest sea animals – the whales. The Great Blue Whale (*Balaenoptera musculus*), the biggest of all, weighs over 100 tonnes when fully grown. Not only are whales the biggest animals in the sea – they are the biggest that have ever lived on our planet, and only the oceans can provide enough food to support them.

Where life began

All life began in salt water. The oceans were teeming with life long before the first plants and animals stepped out on to land, and they have kept their lead ever since.

Top left:
The largest living marine animals (Blue Whales) are over twenty times heavier than the largest living land animals (African Elephants). Despite their enormous size, they are almost weightless in the sea.

Above left and right:
Beaked Whales have snouts elongated into beaks. They have only one or two pairs of teeth and therefore swallow their prey whole. They are amongst the many whales hunted as a food source.

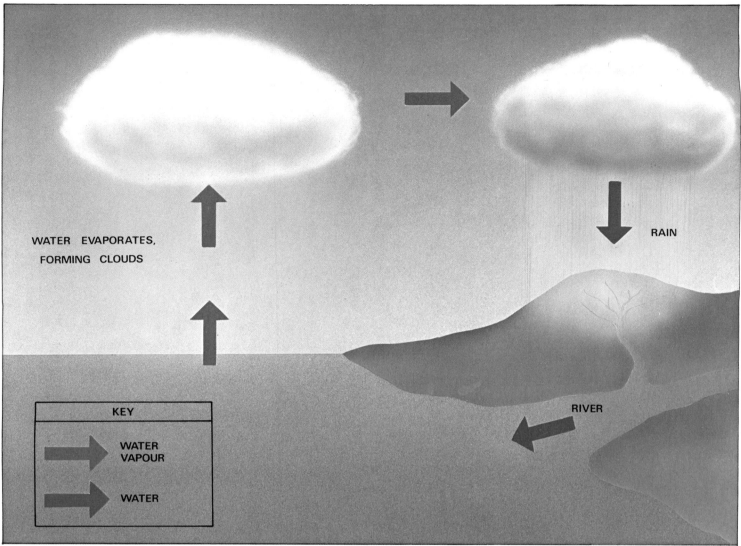

WATER EVAPORATES,
FORMING CLOUDS

RAIN

RIVER

KEY

WATER
VAPOUR

WATER

Above:
The rain cycle. When the sun strikes the sea, warm, moisture-laden air rises and cools to form clouds. Drifted by winds, the clouds move over the land where further cooling turns their moisture to rain. Rainfall erodes the rocks and percolates through soils. Gathering in lakes, streams and rivers, it eventually returns to the sea.

Our own ancestors came out of the oceans long ago, like those of every other land-living plant and animal. Through the age-long history of the Earth, mountains have risen and eroded, whole continents have broken up and moved, but the oceans have gone on and on.

Their water has circulated constantly, evaporating to form clouds, falling as rain, running in streams and rivers, and washing eventually back into the sea with quantities of dissolved salts, sand and mud. That is why the oceans are salty, and are becoming more so as time goes on.

Even the ocean basins have changed. The Pacific Ocean was a different shape 50 million years ago, when the first whales were swimming in it. The Atlantic Ocean is broadening all the time, especially in the south; the Mediterranean Sea of today is all that is left of a much bigger ocean that once spanned the tropics from the Americas to South-east Asia. But there have always been great oceans interconnecting across the world, and they have supported living creatures of many different kinds for 3,000 million years or more.

The plants and animals in the oceans today are a legacy of that long history. In the first chapter of this book we show how life began in the sea, and how the sea has produced many different patterns of plants and animals to make the best use of its resources. It can be seen, too, how these organisms relate to each other in complex food webs, some (the plants) absorbing energy from the sun and materials from the water to build themselves up, while others (the animals) feed on the plants and on each other, to obtain the energy and materials for their own purposes.

In the next three chapters different habitats or living spaces at the ocean edge are described to show how plants and animals live on the shore, in shallow seas, and on coral reefs.

The next chapter deals with the open oceans: the world of the plankton, and of the many kinds of larger animals – fishes, turtles, birds, seals and whales – that live in surface waters far from land.

Then we go on to explore the deep waters of the ocean floor, seeing what kinds of creatures live in the constant darkness, cold and monotony of the deepest corners of the oceans.

The last chapter of all deals with Man's use of ocean life. Man the hunter has always foraged for food along the shores. Some biologists think Man may even have evolved as a shore-living animal, once he left the forests that were his primeval home. For the last few thousand years he has hunted the open sea too, bringing back ever-increasing quantities of fish, shellfish, and other marine foods each year.

Left:
Humpbacked Whale (*Megaptera novaeangliae*). Smaller and less streamlined than Blue Whales, Humpbacked Whales are found in all the world's oceans. Like all other whales they live close to the surface, breathing through nostrils on top of their head: the 'spout' is a jet of warm air, seen when they breathe out.

Below left:
North Atlantic salmon (*Salmo salar*) – one of many kinds of fish that breed in fresh water but live most of their lives in the sea. After travelling the oceans for several years they find their way back to the streams in which they were hatched. Very popular as food fish, many thousands of salmon are netted as they return to the rivers.

Life today

Today our fishing boats and whale-catchers are bringing home sea-foods faster than the seas can replace them. In the last few years we have over-fished stocks of all the common North Sea fish, and many others that are good to eat, in different parts of the world. What was once a cheap and plentiful food for everyone is growing scarcer and more expensive. Wonderful animals – the great whales, for example – are disappearing because too many of them are killed for meat and oil. Pollution, too, is taking its toll of marine life. Oil especially is a menace, but so are many other poisons that we are pouring into the sea from factories and cities. Can Man use the sea wisely, and exploit its resources without destroying them? That is the subject of the final chapter.

The living oceans

Opposite:
Sand is one of the least productive habitats for plants and animals. Shifted by the sea, blown by winds, sharp, abrasive, and poor in nutrients, it gives little protection or support.

Opposite inset:
Alternately covered by the sea and exposed to baking sunshine, this sub-tropical island beach of volcanic sand and shell fragments is virtually lifeless. Old well established dunes (background) often have a covering of hardy grasses and herbs, which bind them together and stabilize them.

The oceans of the world teem with living things. There is far more life in the seas than in all the land habitats (living places) – forests, plains, moorlands, gardens and everything else – put together. However, marine life is not always clearly visible at first glance. You can walk for hours along a sandy beach by the sea without seeing a single plant or animal. It is possible to sail for days across the open ocean, especially in the tropics, without seeing more than a few flying fish and some lone seabirds: but the living creatures are there by the thousand. To see them at home, to discover them properly and find out how they live, we have to move down the beach and plunge into the chilly, unfamiliar world of the ocean.

For land-bound humans it is a wet and uncomfortable world. We cannot see, hear or feel things properly in it, or breathe without tanks of air, or even stay down without draping ourselves with extra weights. But this is the world of the marine plants and animals. It is a lively world, once the diver is used to it, with plenty of action.

Life on the shores

We come closest to the world of the ocean along the shore. Sandy beaches are poor places for plants or animals to live; the shifting sand gives them no foundation or protection, and food is scarce. Pebble beaches have even less to offer. But wander down almost any beach as the tide goes out, to the permanently wet, muddy flats often seen at low water. Here the sand is more stable and there is always a film of water above it. This kind of beach is usually crowded with life.

In each small patch there will be hundreds of tiny snails, dozens of worm casts (the worms themselves are hidden deep underneath), and tiny, shrimp-like creatures in the pools and hollows. Dig down into the sand, and you may find closely-packed cockles and other bivalved (two-shelled) molluscs. (Molluscs are soft-bodied animals, such as slugs, with a head and a foot, and sometimes with a shell.) There will certainly be razor-shells and fragments of many other kinds of shells, showing what other kinds of life lie just below the surface.

Sand-flats like this attract big flocks of shorebirds – oystercatchers, curlews, dunlins, greenshanks and many other kinds – that wade in the shallow water and dig with their powerful bills. The hidden worms and cockles are their food. Enormous flocks of thousands of birds often gather in winter. Their presence is an indication of how rich the sand-flats must be, for thousands of birds need tens or hundreds of thousands of smaller animals to feed on.

Walk on to the nearest rocky outcrop that lies between high and low tide, and there a different pattern of living creatures is seen. There may be red, green and brown seaweeds, often zoned in bands on the rocks, with whelks, barnacles, limpets and other hard-shelled animals growing among them. There may be rock pools, too, encrusted with pinks and containing a population of shrimps, small fish, and tiny red sea anemones. Go further down the shore, wading in the tangle of longer seaweed strands, and there are probably strange coral-like growths, small scuttling crabs, pink, yellow and green sea snails and sponges growing under the rock ledges.

Further down still are prickly sea

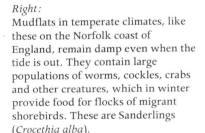

Right:
Mudflats in temperate climates, like these on the Norfolk coast of England, remain damp even when the tide is out. They contain large populations of worms, cockles, crabs and other creatures, which in winter provide food for flocks of migrant shorebirds. These are Sanderlings (*Crocethia alba*).

urchins, larger anemones with candy
stripes, crabs of different colour and form,
starfish and several kinds of spiny, inquisi-
tive rock fish. Comparing rocky shore with
sand-flats, individuals are fewer, but the
variety of species is greater among the
varied surfaces, crevices and pools of the
rock outcrop.

Shore-lines and the sub-littoral zone (that
section of the shore that always lies below
the low-water mark of the spring tides) just
below them, are some of the busiest habitats
of the ocean. They are filled with interesting animals and plants, and we can find and
examine them fairly easily. To see what
lives lower down the seabed, diving gear or
a glass-bottomed boat is needed.

Here is a zone of red and brown seaweeds, that often grow densely enough to
cover the bottom like an underwater shrub-
bery. There are few animals to be seen at
first glance. It is a dimly-lit world even

when the sun is highest, and the animals are
well camouflaged among the rocks and
weeds and shadows. Inshore fishing boats
lay their traps in this zone, taking crabs,
lobsters and crayfish, and netting shrimps
and prawns. Muddy areas nearby often
support rich populations of oysters and
other shellfish, again well hidden from the
sharp eyes of fish and similar predators
that, just as we do, regard them as food.

The open seas

The open seas beyond the shore are no less
well stocked with plants and animals.
Though a cold, stormy stretch of the North
Atlantic Ocean off Britain or Iceland may
seem empty of life, this is in fact one of the
richest seas in the world. If we skim off a
bucketful of its surface waters at the height
of summer, we discover its secret: a wealth
of microscopic plants and animals, swarming in the upper layers. This is the plankton

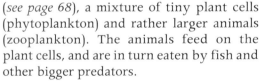

(*see page 68*), a mixture of tiny plant cells (phytoplankton) and rather larger animals (zooplankton). The animals feed on the plant cells, and are in turn eaten by fish and other bigger predators.

If fishing nets of the right mesh-size, at the right season, are set within this layer of plankton just below the waves, the fisherman will catch hundreds of shoaling fish – mackerel and herring, for example – that live at the surface and feed almost entirely on the plankton. Birds, too, know all about the plankton. Gulls, terns, puffins, auks and many kinds of small petrels keep a constant look-out for plankton shoals to feed on; while bigger birds such as gannets feed on the fish that congregate around them. Plankton is also the food of many whales, including the biggest Rorquals.

Life in the depths

Surface waters of the open seas may be rich,
but there is also plenty of life in the depths. A big trawl-net drawn close to the bottom of the same stretch of ocean will catch completely different kinds of fish. They may include Cod, Haddock, Hake: any of a dozen or more species that feed and spend practically all their lives in the cold, deep waters. Tow a dredge along the sea floor, or drop a grab into the bottom mud, and you will find yet another set of animals – the ones that the deep-water fish feed on. There will be sponges, star-fish and brittle-stars, ribbon worms, sea cucumbers, sponges, corals, and a host of other creatures that live on the seabed or buried within it.

The reason why

Why are the seas so full of life? To answer this question we must go back a long way in the history of the world, to the time when life began. Nobody is sure just when this was or how it happened, but everything we know about living organisms suggests that life itself began some 3,000 million years ago, at the surface of a warm sea that contained plenty of dissolved salts and

Top:
Lesser Black-backed Gulls (*Larus fusca*) feeding at sea. Gulls are versatile birds, equally at home on land or ocean. Here they feed on a rich patch of plankton over 160 kilometres (100 miles) from land.

Above:
Often vividly coloured and patterned, animals of the sea floor are no less varied or numerous than those living at the surface. Here a sea slug has similar colouring to the shell of a sea snail, probably to camouflage the sea slug from predators.

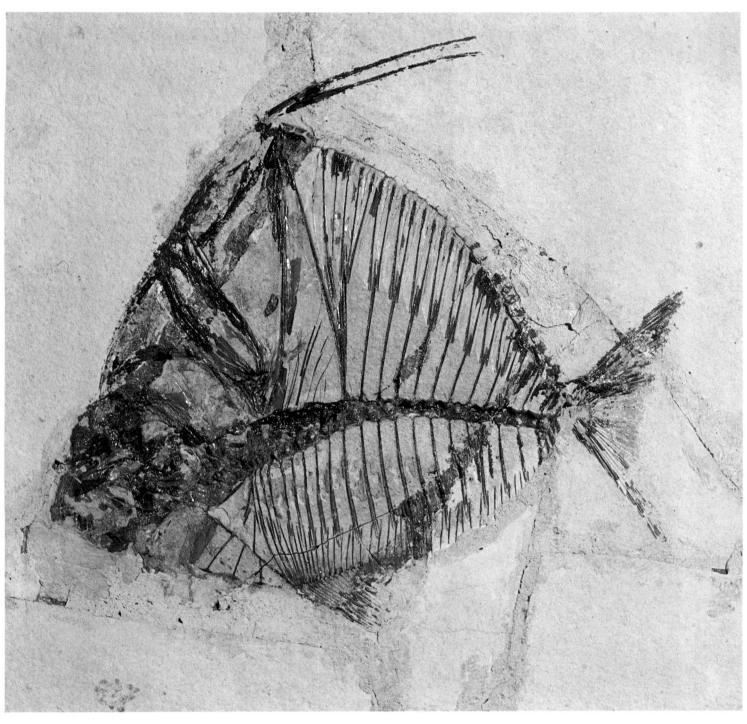

Above:
Most of our knowledge of early forms of life is gained from their fossil remains in the rocks. Worms and similar soft-bodied animals leave very few traces when they die. Vertebrates, such as this fish, leave a precise replica of their bony skeleton, from which we can trace their history and evolution.

gases. Somehow energy was injected into the surface waters, perhaps from sunlight, flashes of lightning or some other source. Using this energy, the molecules of water, salts and gases reshuffled themselves to form primitive, self-sustaining cells – units that were able to draw in more materials from the sea around them, to grow, divide and grow again.

This process must have happened many millions of times before the first truly successful living creatures came into being. But from such simple units developed the whole of life as we know it today. The process began in water, and life continued in water for a very long time. It was probably another 2,000 million years before the first living creatures emerged to colonize the land. We know this from the record of fossils in the rocks. All the earliest fossil traces are of soft-bodied organisms without skeletons: bacteria, algae, jellyfish, worms,

and simple, flimsy plants; or more complex creatures with gills, that could have lived only in water.

At some intermediate time (it is impossible to say when, but perhaps 800 to 1,000 million years ago) life spread from the seas into rivers, streams and lakes. Only in the last 400 million years – about one-seventh of the total span of life on Earth – have plants and animals diversified and spread to cover the land as well. Marine organisms have therefore had much longer to evolve than those of either freshwater or land. It is not really surprising then that the oceans contain so wide a diversity of living creatures, or that so many primitive and ancestral forms are found in the sea.

If the time scale of millions and thousands of millions of years is confusing, think of the whole time span of life (about 3,000 million years) as if it were compressed into a single year. Imagine that the first living cells took shape on New Year's Day. On this scale, life developed and diversified in the oceans through April to July, spreading into freshwater in late August or early September. Through all this time, the land remained a desert. It was October before the first plants and animals emerged – probably from freshwater swamps – to colonize the land, perhaps at the end of October or in early November. Meanwhile, of course, living organisms continued to evolve in the sea, so that new kinds of plants and animals were appearing all the time in the oceans, just as they were on land.

We can use this one-year calendar of events to see when some of the more familiar plants and animals of the ocean first appeared. Bacteria and many kinds of thread-like algae appear as barely-recognizable traces in the rocks at the beginning of life. The green, red and brown seaweeds that line our shores today are much more recent creations, first appearing 400 to 500 million years ago (from early to mid-November on our calendar). The tiny single-celled plants called diatoms and dinoflagellates, that are so important in the phytoplankton, are even younger. They did not appear in their present form until 200 million years ago (early to mid-December), though there were probably other, less efficient plants with a similar role before them.

The earliest sponges and arthropods (animals with jointed limbs, like crabs and lobsters) first appeared over 600 million years ago (about mid-October). Sea snails, bivalved molluscs and cephalopods (the group of tentacled molluscs that includes octopus, squid and the pearly-shelled *Nautilus*) appeared more than 400 million years ago (early November), and the earliest fish date from the same period. However, fish such as our modern sharks and the whole range of bony-finned fishes took almost 100 million years longer to spread into all the oceans (mid-November).

Two hundred million years ago (early December) the oceans were dominated by huge fish-eating reptiles of the Ichthyosaur and Plesiosaur patterns. The first marine turtles, which looked surprisingly like those of today, were also present. Seabirds appeared for the first time about 150 million years ago (mid-December), and whales, seals and other marine mammals replaced the big reptiles some 90 million years ago (mid- to late December). It was a long time after all this – perhaps 2 to 4 million years ago – that Man first set foot on the shore. In our one-year calendar he shows up late in the afternoon of December 31, just a few hours before midnight – the present.

How organisms survive

We have said that life in the sea does not suit all animals, but it is obviously to the advantage of a great many creatures of widely differing shapes, sizes and habits. What, then, has the sea to offer? For the

Below:
Here the whole span of life on earth – 3000 million years – is compressed into a single year. There are very few fossil remains to tell us what was happening during the first four-fifths of the year: most of the record is concentrated in the time from mid-October onward – the last 600 million years.

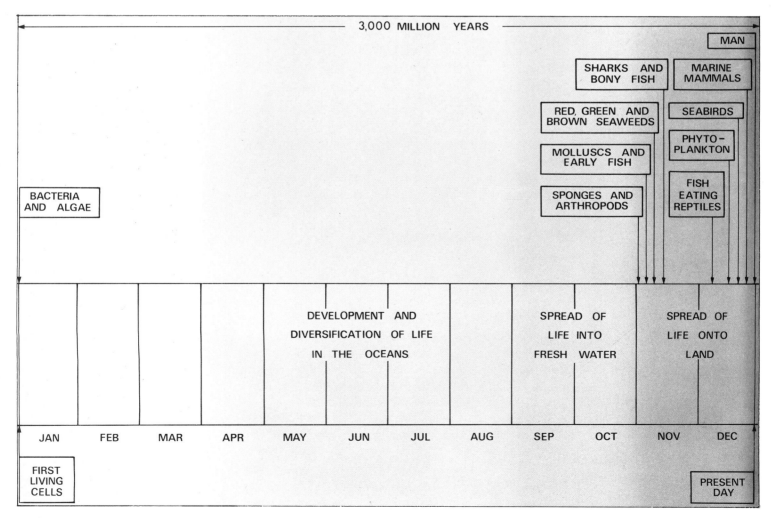

relatively simple plants and animals that make up the bulk of the marine population, much of the answer lies in the content of the water itself.

Sea water is a solution containing most of the minerals of the Earth's crust, with oxygen and carbon dioxide dissolved in it too. It is quite easy to make artificial sea water by shaking a level tablespoonful (28 grams or one ounce) of common salt in a litre of water (just less than two pints). The shaking helps to dissolve gases in the solution. Add half a teaspoonful (three to four grams or 0·15 ounce) of magnesium chloride, and the mixture is improved if you can add a pinch each of magnesium sulphate, calcium sulphate, potassium sulphate, calcium carbonate and magnesium bromide.

This mix contains all the basic ingredients of sea water in roughly the right proportions. Marine plants and animals will live comfortably in it for some time, though it lacks traces of a dozen or more other elements that real sea water has acquired over the centuries — elements that most sea creatures need in tiny amounts to keep them healthy and flourishing.

The same combination of salts is found in sea water all the world over, from the Equator to polar seas, from surface waters to the deeps. Only the concentration varies. On average, sea water contains about 3·5 per cent of salts; marine biologists call this 35 parts per thousand.

In very hot climates, where evaporation is high and rainfall low, salinity at the surface increases. The Red Sea often has salinities of 40 parts per thousand and over. In coastal waters where big rivers empty into the sea — some sectors of the Arctic Ocean, for example — salinity can fall to 30 parts per thousand or less, and to 7 parts per thousand in enclosed seas like the Baltic. But high salinity or low, the relative proportions of the elements remain similar.

The same chemicals in very similar proportions are found in the blood, body fluids and cell fluids of many marine plants and animals, though in greater dilution. This is not surprising when we remember that life began in the sea. It is tempting to think that the body fluids represent the seas as they were in those far-off days, before they had accumulated all the salts that are in them today. But this similarity of chemistry of body fluids and sea water makes it easy for these animals to live in the sea, almost as though they were part of it. They can readily take dissolved oxygen in solution from the sea into their own tissues for respiration, and just as readily get rid of their waste products: carbon dioxide for example, and other substances that are poisonous if not excreted.

Sea water is a great protector of small organisms. It absorbs heat without much change in temperature, and so provides a constant temperature background; plants and animals living in it are never subject to

sudden heating or chilling. It buffers them against jolting and shock, and supports them so that they need very little internal skeleton.

A jelly-fish two metres (six feet) across can manage without internal supports; a land animal that size without a skeleton would flop.

Sea plants have no need for roots or special 'vascular' systems of internal tubing. They can absorb all the nutrients they need directly through their cell walls. It is

the single-celled plants, and the very simple non-vascular algae (seaweeds) that do best in the sea. They can grow big. Some of the giant 'kelp' seaweeds may be 20 or 30 metres (65–100 feet) long – but they still manage without woody skeletons or complicated 'plumbing'.

Similarly, it is the relatively simple invertebrate animals that form the majority of most marine populations. Their basic structures, worked out some hundreds of millions of years ago, have changed very

Above:
Jelly-fish (*Rhizostoma* species) stranded on a beach. Lacking an internal skeleton, jellyfish are normally supported by sea water and they can extend their tentacles and streamers to catch small fish and other prey. On the beach they have no support, and are helpless until the tide comes up again.

Left:
Underwater forest of giant kelp, with a predatory fish waiting to catch its prey. Kelps are big seaweeds, often 20 to 30 metres (65 to 100 feet) long, that grow in shallow waters offshore. Their fronds reach up to the surface from root-like 'holdfasts' on the sea bed.

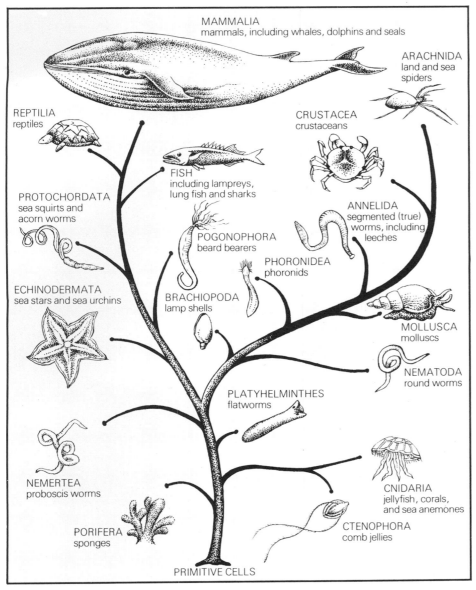

little since their ancestors first appeared in the fossil record. On this page we show a family tree of the major groups of marine animals. The different patterns of invertebrates (animals without backbones) take up most of the diagram, just as they occupy so many important roles in the sea itself. Among vertebrates, the different kinds of bony fishes far outnumber all the rest put together.

Simplest in structure are the drifting animals – the single-celled protozoans and the larger jelly-fish, comb-jellies and their kin. Theirs is an undemanding life, and their body structure is uncomplicated too. Only slightly advanced on these patterns are some of the sedentary animals of the sea floor – sponges, corals and sea anemones, for example – that live by catching tiny particles of food from the water surrounding them. More complex and progressive are the browsing and grazing animals of the shore and seabed. Snails and limpets are good examples. They have to travel in search of their food, and rasp or grind it in different ways before it can be absorbed.

Most advanced of all are the predators: the active hunters that seek out their prey and capture it. These include many of the larger crustaceans (a group which includes crabs and water fleas), echinoderms (the starfishes, sea urchins, sea cucumbers and sea lilies), fishes and cephalopod molluscs. Their hunting demands highly developed sensory organs, well-coordinated movements, speed and skills, all requiring greater complexity of body structure.

Plant and animal distribution

Though the oceans may at first appear to be a hotch-potch of plant and animal life, there is nothing in the least haphazard about the ways in which plants and animals are distributed. The underlying pattern is simple.

All living organisms draw the energy that maintains them ultimately from the sun. Animals cannot tap solar energy directly. Only plants can do that, through the complex chemical process of photosynthesis. Plants in the sea – the single-celled plants of the plankton, and the more complex ones of the shore and shallow seabed – capture solar energy and store it, finally using it for their own growth and reproduction. Plant-eating animals (herbivores) steal this energy when they browse on the plants, storing it and using it for their own growth and reproduction. Meat-eaters (carnivores) feed on the herbivores and also on each other, stealing the energy again at second or third hand, and using it in turn for their own purposes.

Where communities of plants and animals live close to each other, for example in the plankton, the plants supply practically all of the energy directly to the animals. In other communities – on the sandy shore at

Below:
Periwinkles (*Littorina littorea*) clustered on a rock at low tide. These form part of the complex shore food web: they browse on tiny algae, and are in turn eaten by seabirds, fish and other predators.

Above:
A much-simplified family tree, showing how some of the animals of the sea are related to each other. Sponges, comb-jellies, jelly-fish and some simple worms branched off the main stems in the long-distant past. Echinoderms (represented by the starfish), pro-chordates (acorn worms) and vertebrates (turtles, fish, whales) occupy the left-hand branch of the tree, while the many different kinds of invertebrate animals form the right-hand branch.

low tide, for example – plant cells may form only a small part of the energy source. The rest is organic debris: fragments of plants and animals broken up by wave action and sudden death, but still containing stored energy. Surprising numbers of animals of the seashore and the ocean bed feed on debris of this kind. Many are fitted with special mechanisms for filtering it out of sea water, and for sorting digestible debris from indigestible grains of sand and mud. It is the only direct source of energy for animals of the deep seabed, for plants cannot grow at depths greater than about 150 metres (500 feet), and there are not many of them growing below 100 metres (330 feet).

Debris that escapes the nets and funnels of the filter-feeders is finally broken down by bacteria: organisms which are responsible also for the decay of larger animals, body wastes, and all other organic matter in the sea. Theirs is an important role. Breaking down this rubbish releases key minerals for re-circulation and re-use by other plants and animals. Were it not for the bacteria, life in the sea would have halted soon after it began. In destroying organic matter, they release the last traces of its energy for themselves and keep the seas alive.

So we can think of the plants of the sea – whatever their form and wherever they grow – as the energy-providers of their communities. The animals may be exclusively herbivorous, or carnivorous, or they may feed on both plants and animals. Ultimately they all depend on the efficiency of the plants for their welfare. And behind them all are the bacteria, which ultimately break down both plants and animals, releasing their valuable minerals for re-circulation.

Above:
Limpet (*Patella vulgata*). Like periwinkles and other snails, limpets wander over the rocks on their fleshy, muscular feet (*inset*) rasping at the film of algae with their tongues.
Below:
Food web in a marine community.

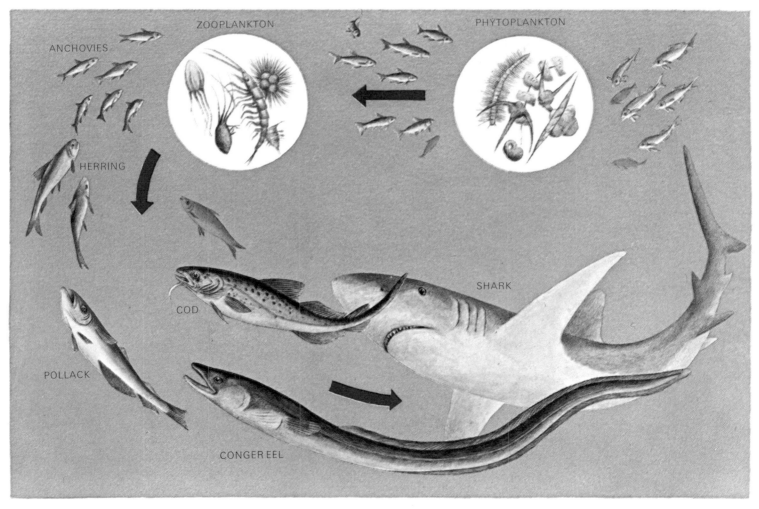

ZOOPLANKTON

PHYTOPLANKTON

ANCHOVIES

HERRING

COD

SHARK

POLLACK

CONGER EEL

Life on the shore

Shores are the one kind of sea habitat that you can examine without going to sea. A boat is not necessary, though it is sometimes helpful to have one, or at least a mask and snorkel. Everywhere in Britain is within quite easy reach of the sea, so practically everyone has seen some part of shore or estuary. In Europe or America the nearest shore may be far away, but it is still the most accessible part of the sea.

Shore habitats vary a good deal in the number and kinds of plants and animals they support. Some are immensely rich: sheltered rocky shores and broad mud flats are often the richest, with dozens or even hundreds of tiny plants and animals to quite small areas. Shingle and sandy shores are poorer: shifting with every tide, they are usually too unstable to attract or pro-

vide homes for living creatures, and some are practically deserts.

The shore line

It is possible to walk some distance along a sandy beach, rocky shore or estuary at high tide and see very little plant or animal life. Salt spray from the sea tends to kill land plants close to the shore, so the trees, shrubs and grasses along the water-line often look withered and beaten. But walk the same way when the tide is out, and you may well see how rich a shore-line can be. The seashore we are looking at here is the zone between extreme high tide (when the sea washes as high as it can) and extreme low tide. Some parts of the zone – those right at the top – are covered by water only at the highest spring tides – perhaps for just a few

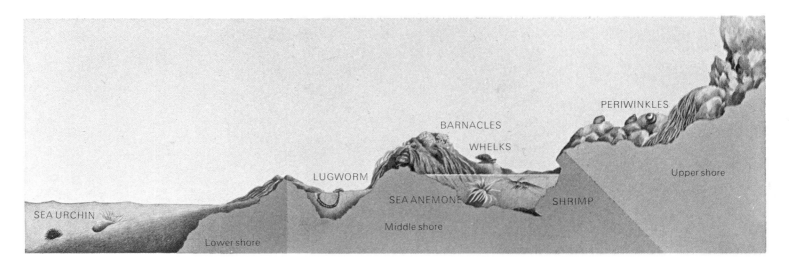

Above:
Zones of the shore. Tidal movements uncover the shore twice daily for several hours. The exposed plants and animals must be able to withstand long periods out of water, sometimes under hot sun or in extreme cold.

Left:
Shore crab (*Grapsis* species) feeding on the jelly of a Portuguese Man-of-War (*Physalia*) washed up among the rocks. Flotsam from the sea provides good pickings for shore-living creatures.

hours every month. But most of the zone is covered and uncovered alternately for periods of several hours each day – twice a day, in fact, for most shores have two high and two low tides in every twenty-four hour period.

So the plants and animals that live on the shore, in this inter-tidal area, have to be able to survive both in air and in water. In fact, most of them are aquatic – mainly water-living – and are most active when the tide is up and the sea covers them.

Then they have special ways of coping when the tide is out and they are exposed to the air. Barnacles, whelks and limpets close up and stay fast on their rocks when the tide falls. Sea anemones round off into shapeless lumps of jelly, crabs hide under rocks in the dampest and safest places they can find. With the tide in, they all come to life. Barnacles open their caps and wave their arms to catch food; whelks and limpets wander quite actively over the rock, grazing the thin film of plant life that covers them. Sea anemones spread their tentacles to catch tiny particles of food in the water, and crabs and shrimps scuttle about to find what they can among the pebbles.

You can see all this activity at high tide if you wear a face mask and look down into the water. But you can see it even without a mask at low tide in some of the rock pools. These are small patches of shore where water is left behind when the tide falls, and big pools often contain all the animals and plants mentioned, and a great many more. They may even contain seaweeds and fish, and they are well worth looking at if you

really want to know how plants and animals live on the shore.

Biologists who study shore creatures think of the seashore area as three parallel zones: the upper, middle and lower shores. The width of the zones depends on three factors: the tidal range, the slope of the shore, and whether the coast is sheltered or exposed. In the rocky shore illustrated, the upper zone would be about 90 to 100 metres (295–330 feet) wide, the middle and lower zones 40 to 60 metres (130–200 feet) wide. They would all obviously be much narrower if the shore were steeper or the tidal range less, and splash effects would be reduced on a more sheltered shore away from the open sea.

Tidal range varies enormously from one part of the world to another. It is highest on continental shores, especially in deep bays and estuaries; mean tidal range is about three and a half metres (10 feet) at Southampton, but seven metres (23 feet) in the Port of London and 13 metres (43 feet) at Avonmouth on the Severn Estuary. On oceanic islands, mean range is usually much less, and tidal movements are very weak in enclosed seas such as the Mediterranean.

The upper shore

The upper shore starts above the highest tide mark, and extends inland as far as spray is carried by winds from the sea. It is hardly a marine zone at all, but we are interested in it because a few very hardy species of marine animals have come up from the sea to colonize it, and dead or

Above:
Sea Slater, a close relative of the common woodlouse or slater, browses among Channelled Wrack on the upper shore.

Right:
Rock Pipit, a small song-bird that feeds on the shore, taking periwinkles, shrimps, sandhoppers, and tiny planktonic animals washed up in the surf.

dying sea creatures from the sub-littoral are often washed up during storms. One of the colonists is the Shore Periwinkle (*Littorina neritoides*), a little blue-back snail with pointed shell. This roams about on cool, damp days, browsing on the orange and black lichens that often colour the zone. Sea Slaters (*Ligia oceanica*), looking like large, coarse woodlice, feed on rotting debris under stones and driftwood. Like the periwinkles, these come from marine stock. Though air-breathing, they cannot wander far from a damp environment, and are at their best in cool, moist conditions with the sea splashing close to them.

Most of the flora and fauna of this upper shore zone come from the land. Such salt-tolerant, flowering plants as sea kale, sea pinks, and the tough, maritime grasses, are the characteristic vegetation, and the animals include centipedes, spiders, bristle-tails and beetles.

Many of these can tolerate long immersion in sea water. One species of midge, *Clunio marinus*, with wingless females, breeds in brackish rock pools up and down the shore. Its larvae, which feed in the pools, seem to be quite happy in salt water. This is not a rich zone for birds, but one species especially, the Rock Pipit (*Anthus spinoletta*), is often seen feeding on the periwinkles, slaters and insects.

The boundary between upper and middle shore is marked by the strand-line — the zone where floating rubbish of all kinds collects at the highest spring tide. After storms in winter and early spring a lot of this flotsam is made up of seaweeds, torn by wave action from the middle and lower shore and the deeper water beyond. Dead starfish, crab shells, and fragments of other sea animals are often washed up with them from the sub-littoral zone, and oil-sodden bird carcasses are all too likely to appear among the debris.

You may also find mermaid's purses – egg capsules of dogfish and other small sharks and rays, laid in shallow waters offshore and anchored among the weeds by their curling tendrils. The empty egg-masses of whelks are often washed up from the lower shore too. The rotting seaweed attracts several species of flies and beetles, whose larvae feed in it and gradually reduce it to liquid. Sandhoppers (*Talitrus saltator*), also lurk in its damp, dark cavities, baling out in their hundreds when the rising tide begins to stir it.

The middle shore

The middle shore is the true inter-tidal zone – the part of the shore that is regularly washed by the twice-daily rise and fall of the tide. As we have said, tidal movements are complex, and differ from one locality to another, but most shores have two high and two low tides in each period of 24 hours and 50 minutes. Creatures living at mid-tide level spend half their lives completely submerged, and half exposed to the air. Clearly those above this level spend more time exposed, and those below spend more time in the water, the actual periods of exposure and submergence depending on

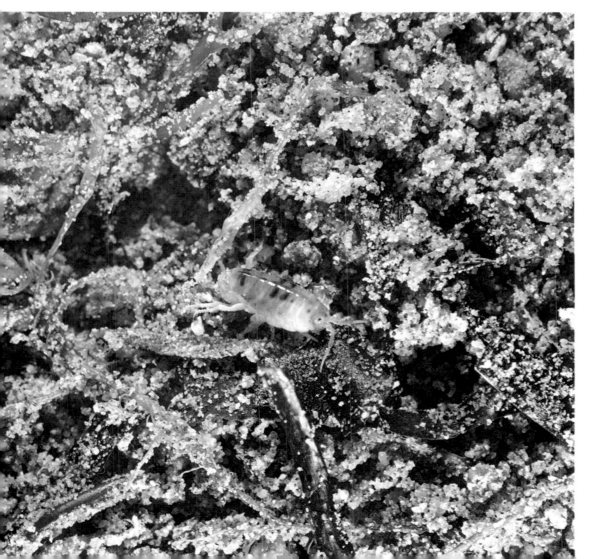

Left:
Sandhoppers live in the damp crevices of rotting kelp, on the strand-line.

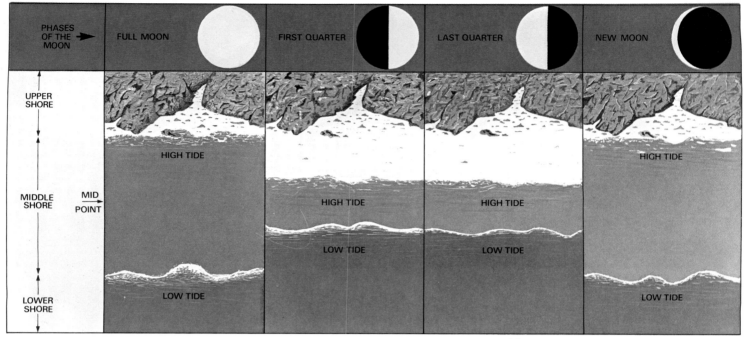

PHASES OF THE MOON →	FULL MOON	FIRST QUARTER	LAST QUARTER	NEW MOON

UPPER SHORE

MIDDLE SHORE — MID POINT

LOWER SHORE

HIGH TIDE / LOW TIDE (across the panels)

Above:
Tidal rise and fall in relation to phases of the moon.

Below:
Zonation on a sheltered rocky shore. With the tide half in, the lowest zones of brown seaweeds are almost covered. Black and yellow lichens mark the upper strand line.

how far up the shore they choose to live.

Apart from the twice-daily cycle of the tides, there is a monthly cycle too, linked with the monthly waxing and waning of the moon. Tidal rise and fall is greatest two or three days after the full moon and again after the new moon: these are the so-called spring tides. The range is least just after the first and last quarters of the moon – the neap tides. At neaps the tide may rise and fall very little above and below the mid-point, while at the springs it may double or treble its range.

So a plant or animal living high in the inter-tidal area may spend several days completely out of water during the neaps period. One living near the low water springs mark may similarly spend days completely submerged, until the tidal cycle works its way round to the springs again. Spring tides tend to be greatest of all at the equinoxes (late March and late September). This is when tidal floods are most likely to occur, especially if strong winds accom-

pany high water and help to increase the tidal range even further.

Normal tides are completely predictable for any part of the coast. We can find when high and low tides will occur, and how high or low they will be, by looking at appropriate tide tables. Wind and waves add complications. Strong onshore winds pile up the water, especially if they are blowing into a bay or estuary, giving higher-than-usual and long-lasting tides. Offshore winds may keep the tides below normal for several days. Waves splash water high up the inter-tidal area, keeping upper-zone organisms moist for longer periods than they could otherwise expect to be. For this reason, the inter-tidal area is broadest on exposed coasts, and narrowest in sheltered corners where wave action is slight.

There is very little pattern to be seen at first glance in the tangle of green, brown and red seaweeds on a rocky shore at low tide. But in fact the weeds are zoned fairly precisely, according to the amount of exposure they can withstand between successive high tides, and many of the shore invertebrates follow a similar zonation scheme.

The key, or indicator, species of algae vary slightly from one British shore to another, and more widely from southern to northern Europe. But inter-tidal zones all over the world show similar patterns of zonation, even when the species involved are completely different from those of Britain.

On a steep rocky shore in southern Britain, orange, yellow and black lichens *Xanthoria parietina*, *Caloplaca marina* and *Verrucaria maura*, mark the lower limit of the upper shore zone. Below them, in sequence down the full length of the middle shore, grows a series of species of green and brown seaweeds, each a zonal indicator.

Highest of all are two or three closely related species of tubular green algae, usually brilliant green and up to half a metre (one and a half feet) long. These are the *Enteromorphas*. They live in high pools, and on rocks and wooden piles up and down the shore, reaching their highest level at, or just above extreme high water mark. Remarkably tolerant of fresh water, but equally at home with the sea washing over them, they clearly mark the top of the middle shore.

Next below them grows the first of the brown algae, the Channelled Wrack (*Pelvetia canaliculata*). Usually a short greenish-brown alga with stumpy, ribbed fronds, Channelled Wrack often forms a narrow band in the uppermost metre (three feet) of the middle shore. It is the hardiest of all the brown seaweeds, able to withstand several days' drying out during the neap tides. Next below it grows a broader band of tan-coloured alga with flat fronds up to

Left:
A colourful group of seaweeds in a sheltered gulley at low tide. Pink *Lithothamnion* and *Corallina* encrust the rocks, with green *Codium cementosum* and yellow-brown *Bifurcaria rotunda.* These species can stand very little exposure, and are found only on the lower shore, or in pools at higher levels.

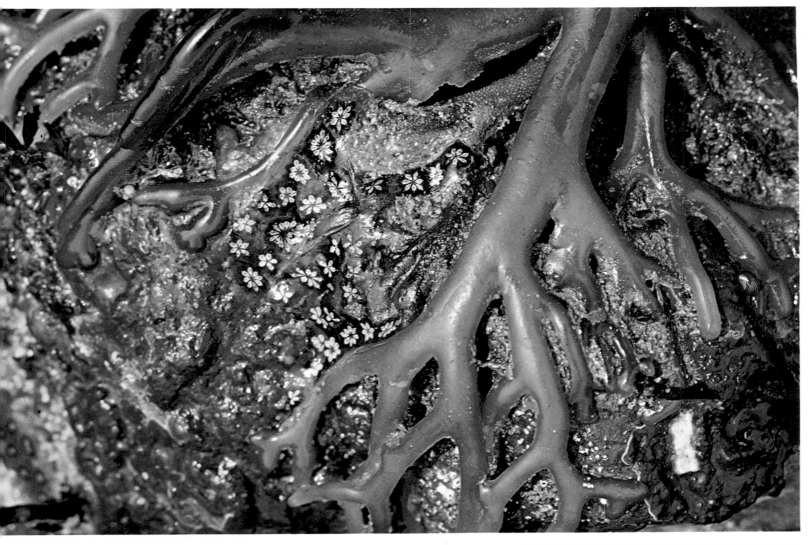

Above:
Holdfast of an oarweed (*Laminaria digitata*) growing at extreme low tide. Many small animals live in and around these knotted 'roots'; the white stars are a colonial ascidian, *Botryllus schlosseri*.

Right:
Bull Kelp (*Darwinia antarctica*) encrusting rocks.

40 centimetres (15 inches) long. This is Flat Wrack (*Fucus spiralis*), a slightly less hardy seaweed that can withstand exposure for 60 to 80 per cent of its time. These two wracks between them mark the upper quarter of the middle shore.

Next below them in sequence comes a broader band of an olive-green alga with tough, rounded stems, growing a metre (three feet) or more in length. The stems carry bladders that help to lift the weed off the rocks as soon as the tide comes in. This is Knotted Wrack (*Ascophyllum nodosum*). It grows thickest in the mid-section of the shore, but cannot stand more than about 60 per cent exposure. Among it or close alongside it is often found the Bladder Wrack (*Fucus vesiculosus*), which seems to like similar conditions, and often jostles for position with the Knotted Wrack. This has flat fronds up to a metre (three feet) long, with single or paired bladders along their length. When the tide is in, the Knotted and Bladder Wracks between them form a dense, swirling thicket, where fish and other hunters dart in search of food.

The lowest quarter of middle shore is occupied by Toothed Wrack (*Fucus serratus*), a shorter brown weed with flat, serrated fronds. This cannot withstand more than about 40 per cent exposure, and so is limited strictly to the wettest zone. On rough days it may never dry out at all. Toothed Wrack grows down to a level below the limit of extreme low water springs, where its place is taken by the much longer, broad-fronded oar-weeds and other algae of the lower shore.

The weeds of the middle shore grow from 'holdfasts' that grasp the rocks firmly. These are not roots; their only task is to provide anchorage. The tough fronds are covered with slime-secreting glands that produce a watery *mucus*. In the sea this helps them to slide over each other without tearing; in air it keeps them from drying out. Mucus may also help to keep at bay some of the other weeds and the animals that are constantly trying to settle on them. Despite this, most of the fronds of the lower middle shore algae are tufted with *epiphytes* – smaller species of algae that live on their surface. Often, too, they are covered with a film of diatoms – unicellular plants that settle on them much as they settle on the bare rock surfaces nearby. This film is a pasture for some of the grazing molluscs to feed on.

Some of the animals of the middle shore are sedentary. Once settled as juveniles, they remain in one spot for the rest of their lives. Others wander freely up and down the shore, especially when the tide is in. But both kinds of animals tend to be zoned, just as the weeds are zoned. The sedentary animals settle in the zone that suits them best, and the mobile return to their appropriate zone – often to an individual favoured home ground, between tides.

Barnacles are among the best zonal indicators of the sedentary forms. Though, as molluscs do, they live in calcareous (chalky) shells, barnacles are in fact crustaceans like shrimps and crabs. Starting life as tiny eggs, they develop in the same way as many other crustaceans, into free-swimming larvae, that live for a time in the plankton. Then the larvae – barely big enough to see – settle in swarms on the rocks and begin to build a fortress of six plates around them. They settle on their heads, leaving their limbs free for kicking. The plates grow as they grow. When the tide is down they rest quietly, sealed in their fortresses by tiny doors. As the water rises and surrounds them, they open the

Below:
Acorn Barnacles exposed at low tide on a shelf of rock. The adults, several years old, have recently been invaded by a swarm of juveniles, some of which will ultimately replace them.

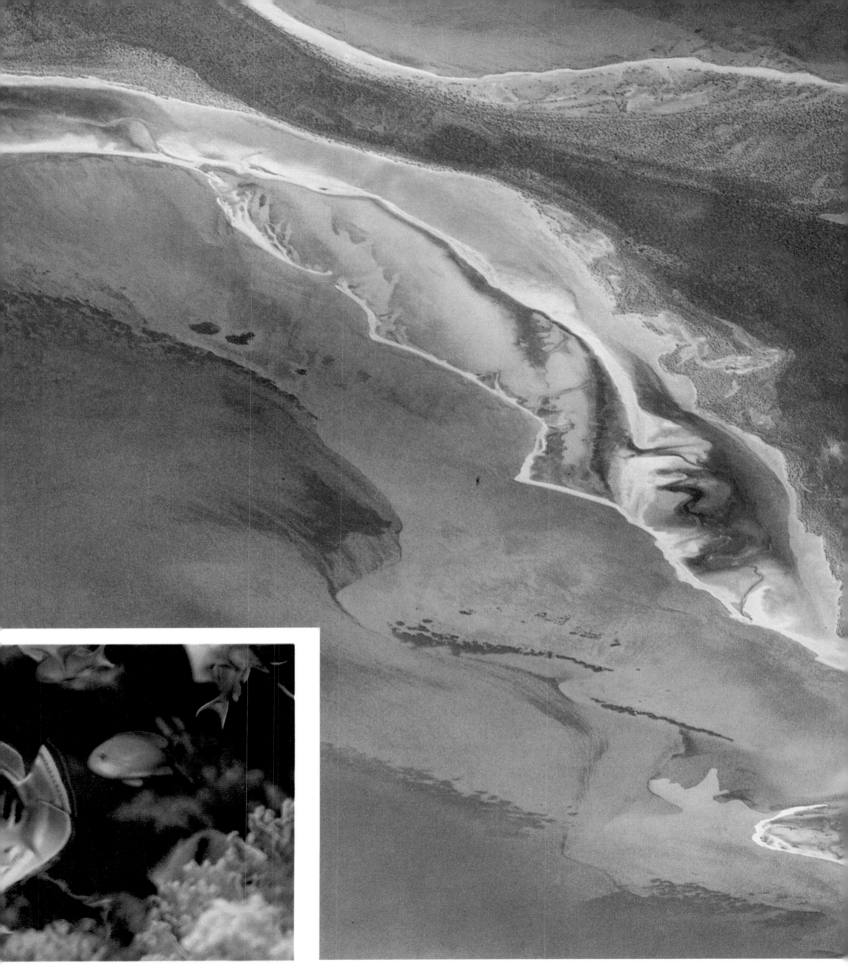

shrubs from the seabed and reach right up to the surface. This is no place for walking; the rough seabed and the tangle of undergrowth make it too difficult. So we swim instead, like all the rest of the larger animals at this level. The tall kelp and the lesser weeds are full of small, darting fish, that browse on the tangle and pick off tiny sea snails and other creatures from the fronds.

Seaweeds grow densely to a depth of about 25 metres (80 feet), then begin to thin out. It is mostly the red ones that are left at this depth, though it's hard to tell what colour they are – the water filters out much of the sunlight and everything looks greeny-blue or grey.

A little further down, and the seaweeds have practically disappeared. Even though the sun is shining brightly overhead, it is as gloomy as a dark forest in here, and we need

Above:
Warming quickly in summer, and well supplied with nutrient salts from the land, shallow seas are usually rich in plant and animal life. They include many of the world's richest fishing grounds.

Inset:
On the seabed of shallow seas, brightly coloured fish can be seen darting amongst the seaweeds.

metres (20 inches) tall but with an immense range of frond-shapes and patterns. Many are familiar as fragments washed up on the beach after storms. Here they are growing in abundance. Brown and green weeds grow among them, but this is predominantly the zone of the red weeds.

Browsing among the weeds of shallow water, and on the substrate from which they grow, are many species rarely seen in the inter-tidal zone. Gastropod molluscs include several shell-less or small-shelled forms. Sea Hares (*Aplysia punctata*), green-grey slug-like molluscs with long tentacles and fin-like flaps on the body, have a small shell almost completely covered by skin. Green Sea Hares (*Elysia viridis*) are shell-less, and browse mostly on green weeds which they match exactly in colour and texture. European Sea Hares seldom grow very long, but American and Australian species can grow much bigger, some reaching the size of real hares and weighing several kilograms.

Sea slugs are superficially similar, but completely without shells. Most are brightly coloured, with decorative appendages and exposed gills on their backs. Sea lemons (*Archidoris pseudoargus*) are lemon-sized and bright yellow, with a warty, lemon-like skin. They feed often on yellow or orange sponges, and are said to take the spicules (chalk needles) of the sponge skeletons into their own tissues, a device that might protect them from being eaten themselves.

Other sea slugs match their prey equally well. *Facelina auriculata*, which feeds on sea anemones, is covered with tufts of appendages that look like anemone tentacles. The appendages may even be armed with cnidaria, for the sea slugs that feed on anemones immobilize the stinging cells with mucus and stow them away in the tips of their appendages.

Other common sea slugs of shallow seas include *Jorunna tomentosa*, a bright orange species, *Dendronotus frondosus*, with a forest of gills on its back, and *Limacia clavigera*, a white or yellowish slug with a row of ten orange-tipped appendages on either flank.

Predators of the shallow seas

Molluscs of a different pattern are among the most active predators of shallow seas. These are the Cephalopods. The 700 or so species alive today are a small remnant of the thousands of species that lived in the past: forms that included the shelled ammonites and belemnites well known to fossil-hunters. We meet some of the few remaining shelled cephalopods in the open seas of the tropics, (*see page 86*). In cool shallow seas we meet three other kinds of cephalopods: cuttlefish and squid, which have a small, hidden shell, and octopus,

a light to see what is happening in the dark recesses. With underwater floodlights all the natural colours come back, and we find ourselves in a colourful world with reds, purples, yellows and blues predominating.

Seaweeds can be found at depths of 40 to 50 metres (130 to 160 feet) in clear northern seas, and almost twice as deep in Mediterranean and tropical waters. Just beyond the surf, dense patches of red and brown weeds grow among the rocks, alternating with bare sand, shingle, gravel or mud. This is a good area to see some of the typical animals of shallow seas.

From low tide mark downward the rocks are often encrusted with pink calcareous algae – red weed-growths heavily impregnated with chalk. Similar growths occur in rock pools higher up the shore. These are species that cannot allow themselves to dry out. Among them grow many other red algae, mostly small – less than 50 centi-

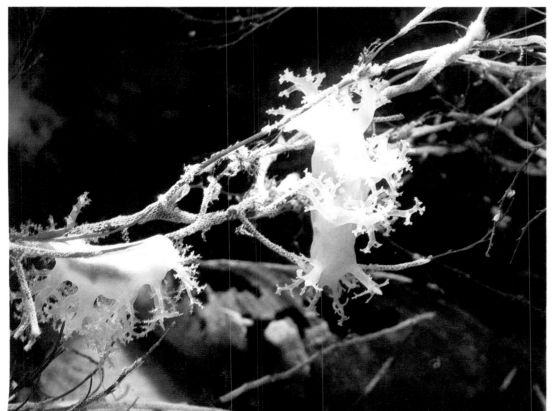

Above:
Sea Lemons are brightly-coloured slug-like browsers of shallow water. Ear-like tentacles mark the front, a crown-like circle of gills the hind end.

Left:
In the Fronded Nudibranch (*Dendronotus frondosus*) branched gills form two rows along the back – excellent camouflage for a sea slug that spends its life among branching fronds of seaweed.

Right:
Lesser Octopus (*Eledone cirrhosa*) watching for prey on a sandy patch of sea-bed. Octopods stalk and lie in wait, flicking out their suckered tentacles to catch fish, crabs and other small prey.

Below
With eyes like glowing headlights, a Common Cuttlefish advances over the rocks in shallow water. Waves of colour shimmer over its skin, helping to hide its movement from unsuspecting prey.

which are shell-less. Nearly all the cephalopods have a solid, rather gelatinous body terminating in eight sucker-encrusted arms. In addition, cuttlefish and squid have two long tentacles, also armed with suckers, that help them to catch and control prey. Because they have 10 arms, they are called decapods, 'deca' meaning 10 and 'pod' meaning limb. Most species have prominent eyes, curiously like the eyes of mammals and birds at first glance, and a sharp parrot-beak buried at the base of the arms.

Octopus and cuttlefish live on or near the sea floor. Species of temperate seas are seldom more than half a metre (18 inches) long and usually much smaller. Masters of colour-change, they blend remarkably with any natural background. Their secret is tiny bags of pigment – chromatophores – in the skin, that change shape and prominence as the position and mood of the animal changes. Octopus (*Eledone cirrosa* is a common European species) crawl along the seabed, cuttlefish (*Sepia officinalis*) move with undulating fins. Both can use a form of jet propulsion, provided by a stream of water that shoots from the mantle cavity (containing the gills) through a funnel on the underside. Squid are more likely to be found in middle and upper waters, where they swim in shoals. They too can change colour rapidly, making themselves almost invisible in the murky depths where they do most of their hunting.

All cephalopods are active predators. Their eyes seem especially sensitive to movement, and moving prey is the only kind that interests them. Octopus and cuttlefish creep along the sea floor, or cast themselves adrift to float slowly until they are within striking distance. Then octopus leap and engulf their prey; cuttlefish shoot out their suckered and spiny tentacles. The prey – usually crabs, fish, shrimps or polychaete worms – are narcotized (stupefied) with venomous saliva and torn apart by the arms and sharp beak. Squid (*Loligo forbesi* is the best-known European species) swim up behind their prey or lie in wait among the rocks and weeds, using tentacles and arms when they are within reach.

Cephalopods court, sometimes with rituals involving fast colour-changes and 'blushing'. Males transfer a package of sperms to the females, which then lay fertile eggs. Usually produced on the seabed, the eggs hatch, and the larvae make their way up into the plankton, where they spend several months as recognizable miniatures of their own kind.

The small cephalopods of European and North American waters are harmless to Man. Larger octopus of warmer waters (for example *Octopus vulgaris*, common in the Mediterranean and sometimes found off southern Britain) may reach two metres (six feet) across and can be dangerous to unwary swimmers. All cephalopods carry a sac of inky fluid (the original of artists' 'sepia') which they release in a cloud when attacked by a predator.

The weed-dwellers

Weed beds on the sea floor are the home of

many kinds of crustaceans. Mantis Shrimps (*Squilla mantis*) and slaters of several species (*Idotea baltica, I. granulosa*) browse in the mud and among the weed fronds. Tiny Ghost Shrimps (*Caprella linearis*) cling with their curved claws among colonies of hydroids, and true shrimps and prawns are widespread in clean, sandy or rocky areas.

Common Shrimps (*Crangon vulgaris*) and prawns (*Leander serratus* and *L. squilla*) are very similar decapod crustaceans, closely related to the larger lobsters and *Homarus gammarus*, Norway Lobsters or Scampi (*Nephrops norvegicus*) and Crawfish (*Palinurus vulgaris*), and more distant kin of the squat-lobsters and true crabs (*see page 34*), but unrelated to the decapod molluscs. They walk and swim actively, scavenging for small particles of food of all kinds.

Above:
High-speed cephalopods with torpedo-shaped body, the squids are usually found among the weed beds offshore. This young squid, stranded in a Devon rock pool, will remain there until released by the next high tide.

Below:
Up to 25 centimetres (10 inches) long, Mantis Shrimps live in warm shallow waters of the Mediterranean. The long, branched antennae 'taste' the ground ahead for traces of food.

Shrimps dig themselves into the sand by day and hunt at night. Lobsters too are more active at night, when they emerge from rocky caverns and holes to search for food. The long, whiskery antennae of all these species scan the ground constantly, and may also be sensitive to chemical stimuli in the water.

Encased as they are in stiff calcareous shells, decapod crustaceans grow by shedding their outer husk periodically. The shell splits down the back, and the animal steps neatly out, one stage larger than it was before. For a few days or weeks after shedding, their new shell remains soft. Often they hide or keep out of the way of predators until the calcium salts harden and protect them again. The jointed tail with its fan-like tip is a powerful defence mechanism; a startled lobster or shrimp flicks its fan sharply under its body, pulling the whole animal back to safety in a cloud of sand. Decapods produce many thousands of eggs, often attaching them to their own body until they hatch. The larvae spend several months in the plankton before settling to find a permanent home on the sea floor.

Starfish and sea cucumbers

Starfish and other echinoderms are common animals of shallow seas. Ophiuroids or brittle-stars – delicate, slender-armed cousins of the more chunky starfish – are often found as abundant, tangled masses, with several dozen individuals in one small area. The Common Brittle-star *Ophiothrix fragilis*, usually orange or purple, is among the most colourful of the European species, with arms up to 10 centimetres (four inches) long. Other species are brown, yellow or grey. *Acrocnida brachiata*, with a smaller disc than the Common Brittle-star, may have arms up to 15 centimetres (six inches) long. They live mainly on sand or mud, sometimes burying themselves just beneath the surface. The long arms sweep about for food, trains of tiny feet and hairs trap small particles and carry them towards the mouth. The arms are also used to give mobility, working together in pairs like oars. Brittle-stars are active creatures, capable of moving long distances over rough ground. Like starfish they develop from tiny transparent larvae, that drift for months in surface waters before settling.

Below:
A common inhabitant of shallow seas, the Common Brittle Star sweeps the sand around it with long, spine-fringed arms. Often broken or lost to predators, the arms re-grow readily.

Another kind of echinoderm common in shallow water is the holothurian or sea cucumber. There are many species around the world, but nearly all resemble tough, warty cucumbers, with rows of appendages and suckers along their underside. At the front there is usually a coronet of tentacles surrounding the mouth, and the body often shows some semblance of the symmetry characteristic of echinoderms. Most European species are 15 to 20 centimetres (six to eight inches) long and two to four centimetres (one to one and a half inches) in diameter. Some tropical species reach almost a metre (three feet) in length. Sea cucumbers feed by shovelling mud and detritus into their mouths. A few species may also extend their oral tentacles to trap food floating past in the water. The common species of European waters are mostly brown or grey (*Cucumaria elongata* and many others), and generally found on muddy or sandy bottoms.

Worms

'Worm' is a name we give to any kind of long, thin invertebrate animal, especially if it digs or burrows – like the earthworms we know well. To biologists the 'worms' we find on the lower shore and shallow seabed are more complex. They can belong to any of half a dozen completely different phyla or levels of organization. Simplest of all are the nemertines or ribbon worms (or bootlace worms, as some are called).

Their bodies are unsegmented and often extremely long. It is hard to say how long, because they are almost impossible to measure accurately, but 20 to 30 centimetres (eight to 12 inches) is not unusual for a naturally extended worm, two or three millimetres (about $\frac{1}{10}$ inch) across. One European species (aptly called *Lineus longissimus*) may reach four to five metres (13 to 16 feet), and a single specimen over 50 metres (165 feet) long is on record.

Above:
Common Lugworms are segmented like earthworms, and burrow easily in soft sand. They live in a V-shaped tube, mucus-lined, through which they pump water continuously past their gills – the pink tufts on their under-surface.

Inset:
Lugworms draw sand into their burrows from a thin surface film, rich in algae and other organic particles. The living matter is digested, the sand passes through the worm, to be voided as 'wormcasts' on the surface of the beach.

Lineus ruber, the Common Red Ribbon Worm of European shallow seas, is a typical nemertine 10 to 15 centimetres (four to six inches) long, often found among mud and stones of the weed belt. Its slender brown body, shiny with mucus, is topped by a slightly broader head, with longitudinal slits and tiny black eyespots on either side. *Lineus bilineatus*, from sandy gravels in deeper water, has two prominent white lines down its back; *Tubulanus annulatus*, a longer worm from sandy seabeds, has several white stripes and crossbars – like seaside rock. *Lineus longissimus*, the true 'bootlace', is often found in lobster pots, to which it is attracted by decaying bait. Usually grey, brown or black, it has a striped head with a dozen or more eyespots on either side.

Nemertines are active predators, especially at night, when they uncoil themselves from a safe hiding place (sometimes the tubular burrow of some other animal) and move slowly in search of prey. They glide on a carpet of mucus produced by glands on their undersurface, and feed mainly on other creatures of the seabed. They creep up on sedentary animals, egg masses and other immobile prey. They catch more active creatures by shooting out a long proboscis or tongue, covered with sticky mucus. Nemertines reproduce sexually, producing larvae which develop in the plankton, or asexually by fragmenting their bodies. A nemertine can give rise to a dozen or more offspring simply by breaking

itself into that number of pieces; each piece grows a new head and tail and goes on its way independently.

The commonest worms of shallow water are the segmented polychaetes, closely allied to earthworms and basically similar in body structure. Only a few, however, of the many dozens of European species resemble earthworms in appearance or behaviour. Most polychaetes are far more brightly coloured and ornate, with a wide range of decorative (though highly functional) appendages.

The Common Lugworm (*Arenicola marina*) of sandy shores and offshore sandbanks is a good example of a 'simple' polychaete. Like many other species, lugworms live in mucus-lined burrows, through which they pump water for respiration. They feed by swallowing sand from the surface, digesting out tiny particles of organic matter and discarding the rest as faeces – the 'wormcasts' that are often visible in shallows and at low tide. Only rarely do they leave their burrows, and then only long enough to find another, similar niche in deeper water. They grow to lengths of about 25 centimetres (10 inches), and reproduce by releasing packages of eggs which, fertilized in the water, develop simply into worm-like larvae.

The complexities of other tube-living polychaetes arise both from stronger, more elaborate tubes, and from more subtle methods of feeding. The tubes may be made of horny or parchment-like material, and

reinforced with sand grains, or may even be impregnated with hard calcium salts. Feeding may be improved by sweeping the sand surface with sticky tentacles, or setting up a network of plumed, often highly-coloured cirri (filaments or probes) to catch small particles directly from the water. The results are often spectacular, especially when several dozen of these very beautiful creatures crowd together to form flower-like colonies.

Sand Mason Worms (*Lanice conchilega*), up to 30 centimetres (12 inches) long, live in sand-encrusted tubes which can often be seen sticking up in tiny forests from the sea floor. Peacock Worms (*Sabella pavonina*) produce mud-encrusted tubes, spreading from the apex a brilliant pink and blue floret. *Potamilla reniformis* has a feathery pink head. *Protula tubularia*, living among gravel in white parchment tubes, produces a vivid red circle of tentacles. Serpulid worms, whose tiny straight or spiral calcareous tubes can often be seen on seaweeds and stones, have feathery crowns of many colours. *Bispira volutacornis*, a colonial sabellid ('feather-duster' worms with tubes made of sand or gravel) worm, has a double whorl of white plumes like tiny ostrich feathers. Those circles of cirri appear when

Above:
Emerging from its tube like a delicate flower, the Peacock Worm displays a double coronet of brilliantly-coloured gills. This polychaete worm lives on mud or fine sand. The gills are a feeding net, catching fine particles which are drawn into the central mouth.

Inset:
Bispira volutacornis, another beautiful fanworm. Tiny eyespots on the gills detect changes of light; at a passing shadow the fans are drawn quickly into the safety of the tube.

the worms are feeding, but snap back into the tubes if danger threatens; a sudden current or passing shadow is enough to make every one disappear instantaneously.

Tube-living polychaetes stay at home. There are others – the errant or wandering polychaetes – that rove in search of their food. Ragworms (*Nereis diversicolor*) and the larger, more colourful King Ragworms (*N. virens*), are good examples. Their colour is variable, tending towards gaudy irridescent greens and purples. Each of their many segments carries a pair of lateral appendages, most of which bear stiff bristles. Antennae and tentacular cirri – long, sensitive probes – crown the head, which also carries several pairs of eyes. Ragworms wander freely over sand, mud and rocks of the seabed, searching out smaller living creatures for food. A trunk-like proboscis shoots out, armed with grasping jaws, to impale the prey. Between hunting excur-

Above:
So-called because of the soft furry spines on its back, the Sea Mouse is actually a segmented polychaete worm. The spines grow from protective plates that cover the animal's gills; they help to camouflage it, and make it less tasty for predatory fish.

Right:
Corals grow in shallow water all over the world, but only in warm seas do they form massive reefs. This is a small branching coral, *Lophelia pertusa*, from shallow northern seas.

sions, they retire to holes and burrows. Some species produce nets of mucus across the burrow mouth, catching small particles that drift in with the respiration currents. One nereid, the brown *Nereis fucata*, lives almost entirely in whelk shells in company with Hermit Crabs. It seems to feed mainly on crumbs from the Hermit's diet.

A less orthodox group of polychaetes, the Scale Worms, have some or all of their dorsal surface covered with overlapping scales. *Hermione hystrix*, a brown scale worm five to six centimetres (two to two and a half inches) long and one centimetre (about half an inch) across, is a typical example. Even further removed from the everyday concept of a worm is the Sea Mouse (*Aphrodite aculeata*) 15 to 20 centimetres (six to eight inches) long and up to eight centimetres (three inches) across, a scale worm in which the dorsal scales are covered with a mat of fine bristles.

Scale Worms live on soft sand or mud, often burrowing horizontally in search of animal material to feed on. The dorsal scales overlie and protect their gills as they plough forward through sand. Why the Sea Mouse alone should have an additional covering of furry bristles is not clear. Freshly caught, the bristles are usually full of sand, which may provide useful camouflage. They are also readily shed and an irritant to the skin; a fish taking a bite at a Sea Mouse might find it as difficult to swallow as we would a packet of pins.

Cold-water corals and sponges

Rocks in shallow northern seas are often encrusted with corallines, sponges and other animal growth. True reef-forming corals grow only in tropical waters (*see page 54*). Corallines are similar creatures – that is, polyps which form a calcareous or horny platform for themselves, but never develop into full-scale reefs. They may, however, form a thick white, pink or grey incrustation with cavities in which other animals find secure niches.

A white branching coral *Lophelia pertusa*, with delicate pink polyps, is sometimes found in fishermen's nets off the Atlantic coasts of Britain. Devonshire Cup Coral

Above:
Red Coral was once widespread in the Mediterranean Sea. Popular for making cheap jewellery, it is now relatively rare. The pink stems are covered with white flower-like polyps. Nudibranchs are often to be found feeding on corals.

(*Caryophyllia smithi*) and Scarlet-and-gold Star Coral (*Balanophyllia regia*) are two species of solitary corals that also occur in clear shallow water off Western Europe. Their polyps resemble small, brightly-coloured sea anemones. Red Coral (*Corallium rubrum*), popular for Victorian jewellery, grows only in the Mediterranean and is now rare. In life its rich red stems are dotted with the white feathery polyps that create it.

Lithophyllum and other calcareous red algae add their colour to coralline reefs, and white soft corals, Dead Man's Fingers (*Alcyonium digitatum*) and its allies grow well. Sea Fans (*Eunicella verrucosa*) and Sea Pens (*Pennatula phosphorea*), with their horny, chalk-encrusted skeletons, grow in tangled abundance. In warmer waters these and similar species may build up a layer several centimetres thick, resembling true coral reefs.

Sponges too add a glow of colour. Though the calcareous sponges of northern seas, for example the Purse Sponge (*Grantia compressa*) and the Crown Sponge (*Sycon coronatum*), tend to be drab and colourless, many of the horny and siliceous sponges are bright orange, green or yellow. Essentially communities of single-celled animals built about a system of passages, sponges draw currents of water through themselves, and the individual cells filter out the small particles of organic matter that make up their food. Calcareous, horny or glass-like siliceous skeletal rods support them internally. Some sponges of deeper water take on fantastic and beautiful shapes, despite their basic simplicity of structure.

Sea squirts

Sometimes similar in appearance, but quite different in internal structure, are the sea squirts that often live alongside sponges on rocks of the shallow sea floor. Solitary individuals of several species grow 10 to 15 centimetres (four to six inches) high, looking like old abandoned handbags of crumpled leather; *Ascidia mentula* and *Ascidiella aspersa* are good examples. Others, such as *Ciona intestinalis*, are translucent, and there are smaller transparent species – *Clavelina lepadiformis*, for example – as clear as delicate blown glass. Each sea squirt inhales water through a hole in its tough casing and passes it through a meshwork of gills, extracting oxygen and food particles on the way. Finally the water is pushed out through a second opening, together with waste products and food remains. If the animal is touched, or even crossed by a shadow, it may contract, shooting jets of

Below:
Sea squirts, so called from their habit of shooting out jets of water, form tightly-packed colonial groups on the sea bed. They feed by drawing in sea water and filtering out the tiny particles of plant and animal material.

Opposite, bottom:
With skin patterns that change colour, the Plaice always matches its background. Here lying on fine sand, its spots – tiny sacs of dark liquid in pale skin – are very small. They grow bigger if the fish lies on pebbles, and spread to cover the skin completely on a background of dark rocks or mud.

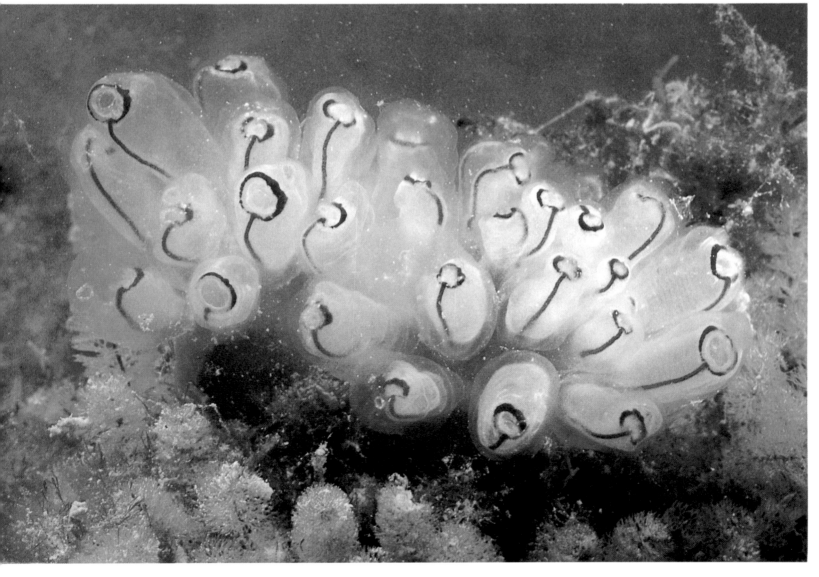

water through both its apertures – hence the name sea squirt.

Many species are tiny and colonial, the individuals living buried in a jelly-like substance which may be brightly coloured pink, blue, green or yellow. The Star Ascidian (*Botryllus schlosseri*) forms rock-encrusting colonies of pink or yellow jelly in which the paler individuals make a pattern of stars (*see page 26*). *Didemnum candidum*, a purple colonial sea squirt, often grows around the holdfasts of laminarian seaweeds in shallow offshore waters.

Fishes in shallow seas

The mass of animal species on the floor of shallow seas provides food for a rich variety of fishes. Narrow seas, like the North Sea, are well stocked with bottom-feeding fish, many of them familiar on the fishmonger's slab. Most clearly adapted for living on the seabed are the flat fish: Plaice (*Pleuronectes platessa*), Sole (*Solea solea*), Flounder (*Platichthys flesus*), Brill (*Scophthalmus rhombus*) and Turbot (*S. maximus*) – all sandy or muddy-hued fish that lie still, partly-buried, and wait for prey to settle near.

Below left:
A colourful assembly of blue and orange sponges, yellow sea squirts and white worm tubes, on the underside of a boulder. In the dim light of a shallow sea bed most of this colour would be lost.

Below right:
The watching eyes of a plaice. As the fish matures and comes to lie on its left side, the left eye moves over to the upper side of the body.

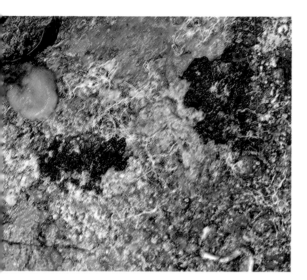

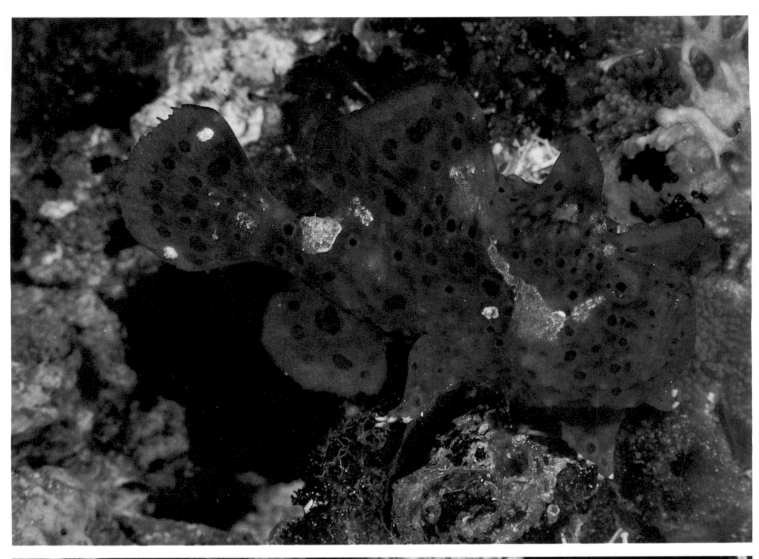

All begin life as normal larvae, perfectly symmetrical, with an eye on either side of the head, swimming freely and feeding on algae and other small organisms in the plankton. After a few weeks of life, they sink to the seabed, and differential growth (more growth on one side than on the other) in the skull brings the two eyes on to the same side. The body flattens, the fins grow to surround it, and the characteristic plate-like shape is complete. In most species the young settle first in shallow water, moving into deeper reaches as they mature.

Angler Fish (*Lophius piscatorius*) with their seaweed-like outgrowths of skin and tatty dorsal fins are almost invisible when lying on a weed-strewn seabed. The curious lure, growing from the upper lip, attracts prey toward the huge, upward-opening mouth. Grey and Red Gurnards (*Eutrigla gurnardus* and *Aspitrigla cuculus*), two of the most colourful European shallow water fish, share their habitat. The three finger-like rays in front of their pelvic fins are sensory organs, used for tasting the seabed. Coralline formations attract many small browsing fish, including the colourful Cuckoo, Ballan and Rainbow Wrasses (*Labrus mixtus*, *L. maculatus* and *Coris julis*), and the Butterfly, Tompot and Striped Blennies (*Blennius ocellaris*, *B. gattorugine* and *B. rouxi*). On sandy and muddy bottoms young Cod (*Gadus morhua*), Saithe (*Pollachius virens*), Pollack (*P. pollachius*) and Whiting (*Trisopterus luscus*) feed avidly, stirring up the seabed and grabbing the creatures they disturb.

Surface waters of shallow seas contain different species of fish, especially the shoaling fish that feed in the plankton. Sprats (*Sprattus sprattus*), Sardines (*Sardina pilchardus*) and Herrings (*Clupea harengus*) are among the best-known; all important commercial fish, their movements and general biology have been studied for many years, and enormous numbers are taken each season as food for mankind. We are not their only predators; young herrings especially fall prey to many other species of fish and to seabirds, and adults are hunted by seals and dolphins as well. But we are probably their main enemy, and the direct cause of their disappearance from many fishing grounds that yielded a rich harvest less than a century ago.

Mackerel (*Scomber scombrus*) and Horse Mackerel (*Trachurus trachurus*) also form huge plankton-feeding shoals, especially in summer when surface waters are especially rich. Mackerel spend early winter on the seabed, feeding on worms, shrimps and other small creatures. In January, they rise to the surface and move offshore to spawn, then spend the summer feeding on copepods (a type of crustacean) and other tiny animals of the zooplankton. In late summer they move inshore, breaking into smaller shoals and feeding mainly on larval fish, which often become plentiful at this time of the year. They too are taken in enormous numbers by Man; mackerel shoals have become especially vulnerable since herrings grew scarcer, and the days of the big mackerel harvests may be numbered.

Opposite, top:
With flattened body and dangling lure, an angler fish (*Antennatius* species) blends perfectly with its background. Smaller fish, attracted by the lure, are drawn into the enormous mouth just below.

Opposite, bottom:
Red-tailed Wrasse or Clownhead, one of the many hundreds of colourful Pacific Ocean coastal fish. Typically hunters of reefs and rocky shores, the Wrasses feed on shellfish and worms, which they dig out and tear to pieces with strong, sharp teeth.

Left:
The sleek, silvery flanks of the Horse Mackerel gleam in the sun. These are surface-living fish, gathering together for safety against predators. They spend most of the year in surface waters, feeding on plankton, but descend to the seabed in early winter.

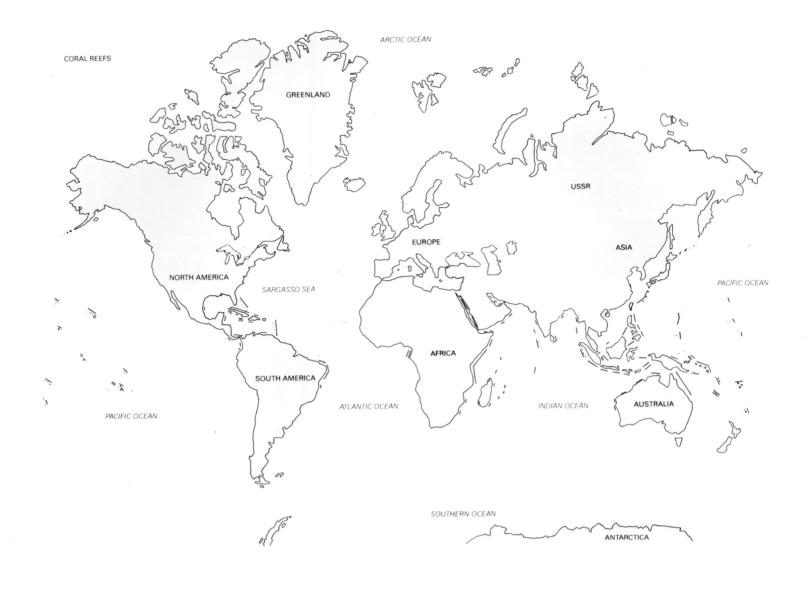

Coral reefs

True coral reefs occur in warm, shallow tropical or sub-tropical seas, where the water is generally clear and fairly calm. They are made of a hard chalky material, usually white or grey, but sometimes tinted pink or delicate purple, that looks at first glance like limestone. In fact it is a kind of limestone, one that is being made all the time by millions of tiny animals – the coral polyps – that live on its surface.

Diversity of coral reefs

Reefs can be smooth and rolling, like broad, open hillsides just below the surface of the sea, dotted with coral rocks and pebbles, seaweeds, and delicate fronds that look like petrified ferns. Or they can take on fantastic shapes, with towers, buttresses and windows, seeming like ruined castles. Some coral reefs are tiny – just patches of coral growing on a rocky shelf in a warm sea. Others are enormous, covering thousands of square kilometres. The Great Barrier Reef off Eastern Australia is almost 2,000 kilometres (1,200 miles) long and over 100 kilometres (60 miles) wide at its broad ends. Many coral reefs are circular or elliptical,

Opposite, top:
Coral reefs of the world. Reef corals grow only in warm, clear shallow water. The biggest reefs of all occur off eastern Australia.

Opposite, bottom:
Different species of corals grow in different ways, contributing characteristic shapes to the underwater scene. This is *Acropora hyacinuthus*, a flat-growing table coral of the tropics.

Left:
Another group of Acroporan corals – the staghorn corals – provides this solid tangle of branches on a Pacific reef. The many crevices and holes are used as refuges by reef-browsing fishes.

Centre:
Several species of calcareous coralline algae, like this *Lithothamnion*, grow among the true corals, filling cracks and adding bulk and strength to the reefs.

Below:
Stages in the formation of coral *atolls* around a volcano. The central cone gradually sinks, leaving the *fringing* reef as a *barrier* reef. This eventually becomes an atoll when the volcano has disappeared completely.

like a ring of clouds round the top of an underwater mountain, or they create a massive breakwater around a tropical island.

Coral polyps appear in all the oceans of the world. We have already met them in the cool waters of shallow temperate seas off Britain and Europe (*see page 49*). They have been found alive and busy in the Antarctic Ocean, and huge banks of cold-water corals are known to exist in depths of 200 metres (650 feet) and more at the edges of the European and the New Zealand continental shelves. But reef-building polyps are more restricted in range. They need sea water of normal salinity, clean and free of mud and other sediments, which in summer reaches temperatures of 21° to 28°C and in winter remains above 16°C.

These requirements restrict reef corals almost entirely to the tropics, particularly to islands, shallow patches of open ocean, and continental coastlines free of muddy estuaries. Coral reefs of many forms are widespread in the tropical Indian and Pacific Oceans on both sides of the Equator; in the Atlantic they are more restricted, occurring mainly in the Caribbean and Gulf of Mexico areas, north of the Equator.

Individual coral polyps are normally less than a centimetre (about half an inch) long. Closely related to sea anemones, they re-semble them both inside and out. But they belong to a division of the Cnidaria called the Scleractinia, all members of which have the special ability to secrete hard skeletons of calcium carbonate crystals in their base plate, and in the internal septa, or curtains, that divide up the body cavity.

In addition, the reef-building polyps are colonial, each individual reproducing by budding to produce mats of polyps linked by living tissues. Where several thousand such polyps grow together, their combined skeletons build up to form hard, stony masses of different characteristic shapes – Brain Corals, Fern and Fan Corals, Stag Antler corals, and many others. Ultimately these individual colonies run together and amalgamate, forming the huge sheets and buttresses of the living reefs.

Coral polyps are the major contributors to their reefs, but they are not the only organisms involved. Bivalve molluscs and barnacles grow among the polyps, their shells adding substantially to the total mass of calcium carbonate as time goes on. Calcareous algae *Lithothamnion* and *Lithophyllum* are important contributors, papering over cracks and filling crevices with solid sheets of stony material. Foraminifera and other minute animals of the plankton (*see page 69*) deposit their shells on the reefs, to be incorporated in time. Corals

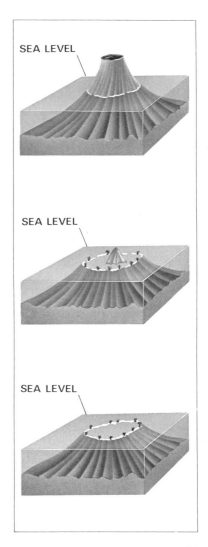

Coral polyps

The coral polyps feed like miniature sea anemones, spreading tiny, threadlike tentacles to catch even tinier particles of plant and animal debris from the water that swirls constantly around them. They feed mostly by night, retreating by day into the tiny calcareous cups they have made for themselves. So a coral reef in bright afternoon light looks relatively dead. Only in the half-light of evening, and by floodlights at night, can the polyps be seen in full activity and in their true, startlingly brilliant colours. A live reef by night, with all its polyps extended, is like a single huge, hungry animal. Its nightly catch of tiny organisms and organic particles must be enormous.

So far as we know, coral polyps are mainly carnivorous, and the food they catch and take in by mouth is their main source of nourishment. But they have another source too – one that helps to explain why coral polyps (which are animals) require bright sunlight to keep them healthy and fast-growing. Their body tissues contain microscopic plant cells called zooxanthellae, which live in a state of symbiosis with the polyps. This means that both the polyps and the zooxanthellae benefit from the relationship. Corals grow best in shallow water under strong sunlight – conditions in which the zooxanthellae might be expected to be photosynthesizing actively. The zooxanthellae probably make use of the polyps' waste products and perhaps supply them with sugars. In

grow quickly. Some authorities consider a centimetre (half an inch) per year a reasonable rate of growth. A sunken ship may become completely encrusted and almost unrecognizable in 20 to 30 years. But it takes many thousands of years to build up a substantial coral reef.

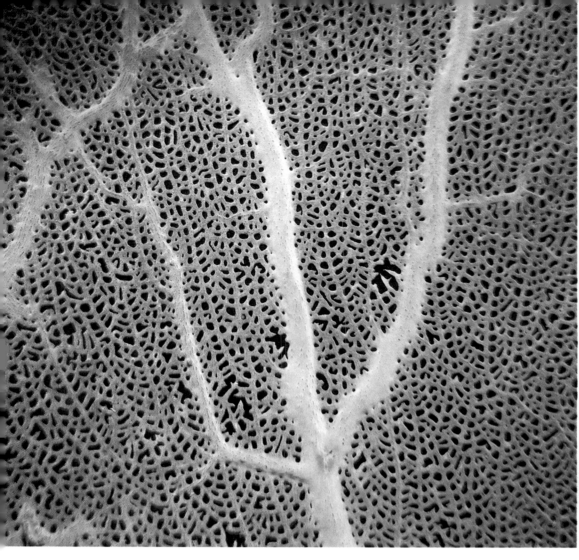

return, they are held at a favourable depth by the polyps, and protected by their defence mechanisms.

Reef-building polyps are active from sea level down to 40 to 50 metres (130 to 160 feet). In the dim light at greater depths, the zooxanthellae are unable to make their contribution, and growth is negligible. Where reef corals occur at greater depths, or indeed above sea level, there can only have been a relative change in the levels of sea and land.

We know of many such changes in the recent history of the world, enough to account for the many different shapes of coral atolls, fringing reefs and cays (small islands). Enormous fossil coral reefs tell us, too, that coral seas once extended over much of Europe, and covered vast stretches of North America.

Reefs as a community

Because of the uneven, patchy growth of dozens of different species of coral, reef surfaces are inevitably rough, with crevices, cavities and deep clefts like an old cliff face. The nooks and crannies of a big reef provide shelter for hundreds of other kinds of marine animals, and anchorage for weeds, sea grasses and other plants. Reefs are community centres, far richer both in species and in numbers of individuals than most other living communities.

This is especially true of Pacific and Indian Ocean reefs, which tend to be richer in variety than those of the Atlantic Ocean.

One large group of corals, the genus *Acropora* that includes the many-branched Elkhorn and Staghorn corals, is represented by 150 species in Indo-Pacific reefs, but only three species in the Caribbean. Among the associated forms there are abundant soft corals (Alcyonarians) and sea fans (Gorgonians) on Atlantic reefs, but almost every group of reef-living invertebrates from flatworms to clams is represented by a greater variety of species in Indian and Pacific Ocean reefs.

Visitors to the Great Barrier Reef, indeed to most of the Pacific island reefs, can walk out on to the coral flats at low tide. Many species of this region can tolerate a few hours' exposure at low spring tides. Other reefs can be examined more comfortably from glass-bottomed boats. It is better still to get down among the turrets and towers of coral with shallow-water diving gear, and see the incredible wealth of life at close quarters.

In the dappled sunlight, 10 to 15 metres (33 to 50 feet) down there is little loss of colour. Everything looks greenish, but the yellows, reds and purples of the plants and animals shine through like brilliant enamels. The seabed, usually covered with fine coral sand, seems curiously clean and empty, especially in comparison with the fine sands and muds of cooler temperate seas. But coral sand is coarse-grained and shifts readily when the sea is disturbed; it is not a safe medium for soft-bodied worms and other delicate creatures to hide in. Like everything else, they live on the towering walls and cliffs of the coral reefs themselves.

A few flat fish, camouflaged by their colouring, lie in wait on the seabed, making an occasional dart at passers-by. Peacock Soles blend with their background; like many other flat fish, they can change the colour patterns of their upper surface to match the sand or shingle on which they are lying. Carpet sharks (*Orectolobus* species) –

curiously flattened dogfish, with an outline fringed by weed-like tufts of skin – look exactly like weed-covered patches of coral. Stone Fish (*Cynanceja verrucosa*), with a rough, reddish-brown upper surface, resemble large rocky pebbles lying among the sand. They have an armoury of venomous spines to discourage potential predators.

Small, bottom-living rays flap inconspicuously along the sand, sucking tiny fish, shrimps and occasional molluscs into their cavernous mouths. They too have sharp spines, jagged like bayonet blades, that slash when their owner leaps to avoid a larger predator.

The 300 or more species of corals that make up the walls and 'bommies' (isolated columns or coral-heads) of the Great Barrier Reef seem at first glance to be scattered at random. In fact they are graded according to exposure, currents and a dozen other factors, just as rigorously as the plants and animals of the inter-tidal zone. Common on the exposed outer surface of the reef are the Acroporans – tough branching corals that interlock to form a hard, resistant wall. On top, and in the more sheltered areas within the reef, smaller branching forms mingle with flat slab corals of many species, including the ubiquitous *Fungias* – like large inverted toadstools – the blunt, chunky *Stylophoras*, lumpy *Milleporas*, and sponge-like *Porites* of several species. In the most sheltered and protected zones grow star corals (*Montastrea*), fragile *Lobo-*

Below:
Wobbegongs or carpet sharks are small sharks, harmless to man, that feed in shallow waters around the Australian coast. This species, *Orectolobus ornatus*, is banded, and covered with fronds – outgrowths of skin – that match the seaweeds.

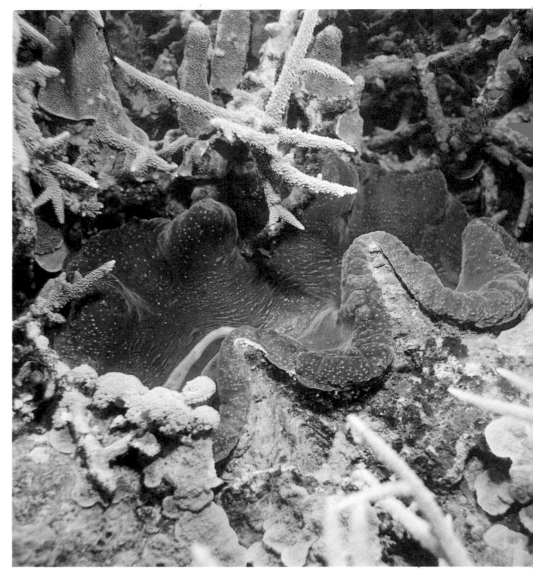

phyllias and delicate pink organ corals (*Tubipora* species). Brain corals too occur in quiet corners away from strong currents and wave action.

Within the sheltered lagoon of a fringing reef, stony corals may find themselves competing for space with many species of soft corals and fan corals, whose structure would not stand up to wave action and currents in more exposed situations. Sponges, bryozoans (colonial animals that produce moss-like encrusting colonies) and vividly coloured anemones add to the confusion of species.

Here and there among the sessile (fixed to the bottom) forms appear the wavy shells and coloured lips of Giant Clams (*Tridacna gigas*), almost completely buried in the coral. Like other bivalved molluscs, these are filter-feeders, drawing in currents of water and sifting out microscopic food particles. The coloured 'mantle' edge –

Above left:
Dozens of species of fishes, of all shapes and sizes, live on and about the coral reefs. This Stonefish (*Synanceja horridus*) looks like a lump of coral rock. Its spines are covered with an agonizing poison.

Above:
Huge bivalved molluscs, Giant Clams feed by filtering sea water. Up to a metre (3·3 feet) across, the two halves of the shell are held together by powerful muscles, that relax when the animal is feeding. The blue mouth edges contain algae, which photosynthesize in sunlight.

Left:
Star coral, one of the.many corals that contribute to the reefs of the Caribbean islands. The individual polyps grow outward, forming a solid mass of calcareous rock at the centre.

often bright blue – houses zooanthellae, which probably benefit the clams as much as they benefit the surrounding coral polyps.

A coral reef is difficult for humans to walk on. Not quite strong enough to bear a man's weight, the fronds of coral break unexpectedly, and the rough edges inflict nasty wounds that often heal slowly. But smaller creatures have no problems of weight, especially when the tide is in. All kinds of mobile predators can be found wandering the reefs, particularly after dark when the coral polyps have emerged from their protective cups. Anyone familiar with the animals of the European shore would recognize many familiar forms, though none of the same species. Most of the reef inhabitants are brighter and more florid; many are larger, and overall they are present in much greater abundance and variety. About 3,000 species have been found on the Great Barrier Reef alone. Well over a third of them are fish.

Brilliantly-hued sea slugs, nudibranchs and shelled gastropod molluscs browse among the gorgonians and sponges, performing much the same roles as their counterparts in cooler seas. Shell collectors would recognize many of the molluscs immediately – the spotted and striped volutes, cones and bubble shells, bailers, turbans, tritons, cowries and abalones are in demand all over the world. What might surprise the collectors is the sheer beauty of the living animals inside the shells.

No less vivid are the shrimps and Hermit Crabs that scavenge among the polyps. And no more astonishing is the way that these lurid, multi-coloured creatures merge with their backgrounds in the brilliantly lit, sun-dappled waters of the reefs.

It would be difficult to imagine four animals more strikingly patterned, or more different from each other, than the Red-banded Coral Shrimp (*Stenopus hispidus*), the orange and black Sea Slug (*Cyrce nigra*), the crimson and gold nudibranch *Hexa-branchus imperialis*, and the brown mottled Tiger Cowrie (*Cypraea tigris*). Yet each has its own kind of vanishing trick to play against the rich tapestry background of the reef. The shrimp, with crimson and white stripes aggressive as a football scarf, blends into a background of red polyps and white coral. The sea slug and the nudibranch disappear among brilliantly-hued sponges and sea fans, and the cowrie is lost against the mottled corals which are its feeding grounds.

Below:
Brilliantly-coloured Boxer Shrimp, so-called from its habit of waving its large pincers when danger threatens. Bright colours are toned-down in shallow water so the shrimps merge with their background.

Tubiculous polychaetes add their delicate, shapely fans and whorls of tentacles to the tapestry, competing with the coral polyps and anemones at their game of catching food. Errant or wandering polychaetes creep and wind among the crevices of the reef surface in a constant search for unwary prey. They too have the magical trick of matching their background, despite brilliant colouring. There are sea squirts in abundance, the colonial species especially adding brilliant splashes of colour to the surfaces they inhabit.

Echinoderms are particularly well represented on the reef. Lumpy sea cucumbers of many species ingest the coral sand, passing it through their digestive system to extract organic material. Sea urchins are plentiful. Some, like the several species of *Diadema*, have slender knitting-needle spines three to five times the length of their body, that move in silent warning if danger – even the shadow of danger – approaches. The spines are tipped with poison, and break readily in the flesh of anything that brushes against them.

Pencil urchins (*Heterocentrotus* species) use their stubby, club-shaped spines to wedge themselves into crevices, making an almost impossible target for predators. Like *Diademas* they browse the coral with their five strong teeth, rasping and grating in

Above right:
Cerise Nudibranch (*Hopkinsia rosacea*), a soft-bodied mollusc that browses in shallow water off the coast of California.

Above left:
Tiger Cowrie (*Cypraea tigris*). In life the spotted shell is part-covered with a fleshy mantle, also brilliantly coloured. Related to whelks and other sea-snails, the cowrie shell has an internal spiral.

Left:
Long-spined sea urchin (*Diadema antillarum*). The spines, hinged at their base, help to keep predators at bay. Small fishes sometimes live among them, taking advantage of the protection they offer.

61

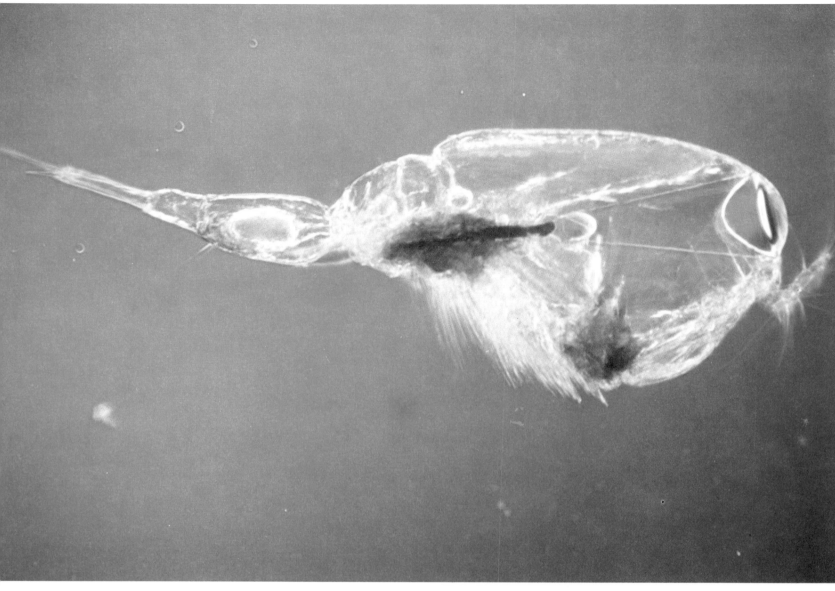

reddish-brown soup. The smallest are often the most plentiful, and ultimately the most important as the starting point of food chains. About three-quarters of the chlorophyll (the substance which is contained in *chloroplasts* and makes plants green) in the oceans may be contained in diatoms and flagellates each less than one-hundredth of a millimetre in diameter.

The smallest animals of the seas

Living in close proximity to the phytoplankton are the hordes of minute animals that make up the zooplankton. There are basically two kinds of zooplanktonic animals: those (like the larvae of sessile shore animals and bottom-living fishes) that spend only part of their life at the surface, and the many other species that are entirely planktonic for the whole of their life cycles. The two kinds live side by side, feeding avidly on the phytoplankton and on each other.

The smallest members are protozoans: single-celled animals that, like diatoms and flagellates, often have highly decorated and beautiful casings. Foraminifera, with bubbly shells of calcite, and siliceous Radiolaria are widespread in all the oceans.

Right:
Sea gooseberry. The transparent spherical body propels itself through the water by eight rows of beating cilia (fine hairs) that beat in unison. Two long trailing tentacles catch small animals and drag them towards a mouth at the end of the body.

Below:
Portuguese Man-of-War, a large siphonophore with a blue or purple float. The long tentacles may trail eight to ten metres (26 to 33 feet) behind. Animals caught by them are hauled up to the mass of polyps beneath the float and killed or immobilized by the stinging cells.

Their cases are pierced by holes through which they stream long sticky threads to trap the phytoplankton. Ciliated protozoans also abound, and there are many flagellates that closely resemble the microscopic plant-flagellates. They lack chloroplasts (where the process of photosynthesis occurs) and feed on bacteria and other organisms in the smallest size range.

The many-celled animals that make up the zooplankton are drawn from almost every phylum or major group in the animal kingdom. Crustaceans are probably the most widely represented; copepods, mysids, euphausiids and many other groups of crustaceans can be found in great variety, ranging in length from one millimetre to several centimetres (less than 0·03 inch to several inches). The smallest tend to be brush-feeders. Their mouth-parts and legs

are covered with bristles which sweep through the water to catch diatoms and other small particles. The larger ones tend to be carnivores, unashamedly laying about them with jaws snapping at their smaller neighbours. But there are many other kinds of animals in the zooplankton, and many other methods of feeding.

Among the larger creatures are several related to sea anemones: the delicate jellyfish, which pulsate in the water and trail their stinging tentacles beneath them, the sea gooseberries and other comb-jellies that propel themselves along by rows of fine cilia, and more elaborate jellyfish called siphonophores – whole colonies of polyps supported either by swimming bells or by gas-filled floats. Like sea anemones, jellyfish (often called medusae) and siphonophores have tentacles armed with poisoned darts, that they use in capturing and subduing prey.

Even a small jellyfish can give a painful sting if you brush against it in the water; a large one (and the biggest reach over a metre in diameter) can cause serious injury or death. Siphonophores too can be dangerous. The Portuguese Man-of-War (*Physalia physalis*), a beautiful purple float with trailing mauve and blue tentacles, has a particularly sharp venom that raises weals on human skin, and can still be effective several days after the animal has died. The 'keeled' float of this creature allows it to sail before the wind, so it moves through the sea trawling for plankton and small fish as it goes. *Velella*, the By-the-wind sailor is a similar but simpler siphonophore with inflated sail. These are mostly creatures of warm seas, but are occasionally washed up on to the colder beaches of Europe.

Several kinds of molluscs live in the plankton. *Glaucus*, an elaborately fringed nudibranch, browses on siphonophores, as many of its bottom-living cousins browse on anemones. *Ianthina*, the Purple Sea Snail, looks just like a sea-going garden snail with a fragile blue or purple shell. It hangs upside down from the surface film of water, suspended by a raft of bubbles encapsulated in toughened slime. It too browses on siphonophores, also taking copepods and other planktonic creatures.

Pteropods, or sea butterflies, are gastropod molluscs, some with small coiled shells and others naked. Their wings, sideways extensions of the foot, beat to keep the animals afloat and moving in search of prey. Thecosome (shelled) pteropods feed on microscopic plants of the phytoplankton, catching them in rows of cilia that carry them like conveyor belts to its mouth. Gymnosome (naked) pteropods are more actively predatory, feeding on small planktonic animals including their own kind.

Arrow worms, or Chaetognaths, are transparent, quill-like animals one to ten centimetres (0·3 to 4 inches) in length, that live almost exclusively in plankton all the world over. Tiny eyes, simple bristle-like jaws, sensitivity to movements in the water nearby, and a dart-like turn of speed are their stock-in-trade. They are among the most lively and active predators of the plankton, darting rapaciously after the larvae and small crustaceans that make up their food.

Polychaete worms too are found in the permanent plankton; *Tomopteris* is a widespread genus. Their lateral 'limbs', or parapodia, are developed into paddles that help to keep the worms afloat as they

Above:
Bubble snails hang from rafts of bubbles that they produce themselves. Inhabitants of tropical seas, they are often found feeding, like this one, on By-the-Wind Sailors.

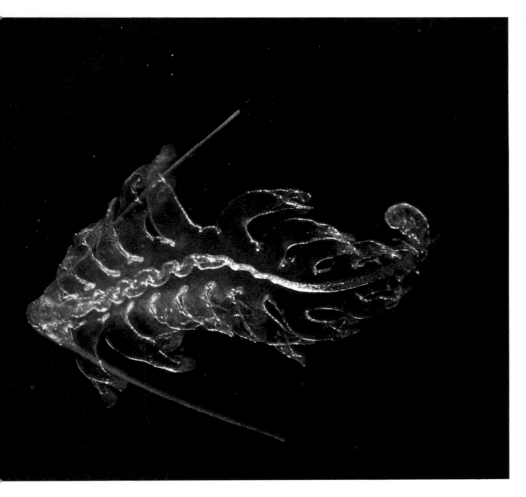

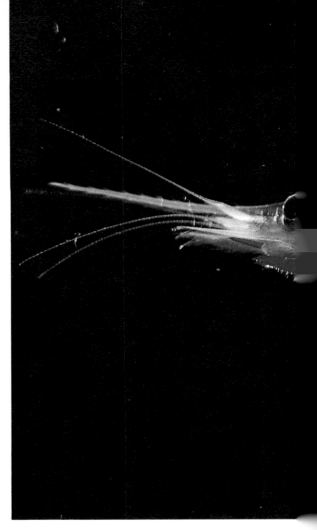

Above:
Tomopteris species, a predatory planktonic worm.

wriggle through the water. 'Palolo worms' are the wriggling tails of bottom-living polychaetes *Eunice viridis*, that break away from the parent bodies and float to the surface of the Pacific Ocean during the neap tides of October and November. Filled with eggs and sperms, but very little else, they burst at the surface in their millions. The eggs, duly fertilized, develop into transparent larvae that eventually sink and settle on the seabed.

Also transparent, though more passive in their lifestyle, are the strange Larvacea and Thaliacea. Looking rather like complex jellyfish, they are related closely to sea squirts, and more distantly (perhaps more surprisingly) to fish and other vertebrates.

Larvacea are tadpole-like creatures a few centimetres long that live in 'houses' of tough transparent jelly. Through open windows in the houses they draw currents of water, filtering-off tiny cells of phytoplankton and catching them in mucus nets. Thaliacea include salps and doliolids – barrel-like creatures of translucent jelly, usually six to eight centimetres (two to three inches) long, and pyrosomids (*Pyrosoma*, for example) which are long, sausage-like colonies of sea squirts adapted for life at the surface. All the Thaliacean are capable of slight movement by jet-propulsion, and feed by filtering the water that passes constantly through them.

The name *Pyrosoma* means 'fiery-body'. Several of these and related species are luminescent, flashing brilliantly when the water around them is disturbed. So are many other creatures of the plankton, notably the flagellates *Noctiluca* (night-light), and *Pyrocystis* (fire-bladder), and several species of sea gooseberry. Their flashes of cold green light may be a defence mechanism, deterring predators that approach and snap at them. But there are times when, in warm, calm seas just after sunset, they light up spontaneously with the movement of the waves. The sea as far as the horizon glows eerily with millions of tiny, flashing lights.

The countless larvae that make up the temporary population of the plankton include representatives from the shore, from shallow waters, and from all other levels of the ocean. With a rich sample of springtime or early summer plankton, a good microscope and a key or reference book, you can identify the larvae of dozens of animals from sponges to crabs and barnacles, from brittle-stars to cuttlefish and polychaete worms. With practice, almost anyone can key them into their major groups, but it may take an expert to identify them down to family, genus and species.

The larvae, newly-hatched from their eggs, feed on the tiniest particles, usually plant cells or protozoans. As they grow, they graduate to slightly larger foods – the smallest multi-cellular zooplankton, including younger forms of their own kind. Growing further, they may move up the food chains to feed on small predators, eventually leaving the dangerous waters of the plankton altogether and taking up life in haunts similar to their parents'.

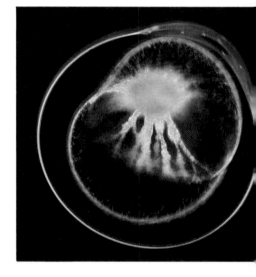

Left:
A hydroid medusa capturing another
hydroid.

Seasonal movements of the plankton

Phytoplankton remains within a few metres of the surface; zooplankton generally shows marked daily and seasonal movements within a range of 200 or 300 metres (650 to 1,000 feet). Seasonal movements are most marked in temperate and polar seas.

During winter, when the sun is low in the sky, light is poor and surface waters are cold, the remnants of summer plankton remain hundreds of metres down, waiting in a quiescent state for the return of the sun.

In early spring, the diatoms and other microscopic plants begin to photosynthesize actively, take in nutrients from the waters around them, and divide repeatedly. Then the zooplankton rises from the depths, soon to be joined by new generations of larvae that increase its bulk a thousandfold. As the sun rises higher each day, the zooplankton begins to oscillate, sinking to depths of 50 to 100 metres (160 to 330 feet) across the middle of the day, and returning to feed avidly among the phytoplankton each night.

The reason for this movement is uncertain. It may help to protect the larger animals of the plankton (the so-called macroplankton) from the creatures that hunt them, though many of the hunters – shoaling fish such as herrings and mackerel, for example – move up and down with the plankton too. It may keep them away from the plants at the time of day when photosynthesis is strongest, and waste products from the plant cells taint the sea.

Surface dwelling creatures

Plankton is fair game for all the larger animals of surface waters, from the smallest fishes to the largest whales. Collectively these swimming creatures of the surface are called nekton. At the smaller end of the scale they include surface-*living* euphausiid 'shrimps', for example, *Meganyctiphanes norvegica* of northern waters and *Euphausia superba* of the far south – both shrimp-like crustaceans four to ten centimetres (one and a half to four inches) long that feed on microscopic algae.

At the height of summer, these and other

Below:
Frigate Bird in flight.

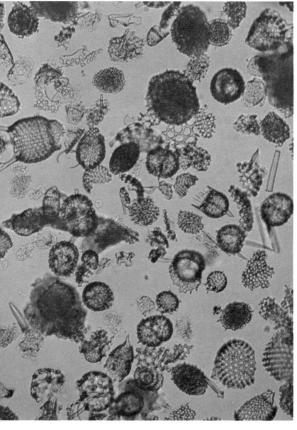

related species form enormous shoals several hectares across, known to whalers and fishermen as 'krill'. They are an important food for whales. Next largest, are the many species of shoaling fishes that feed on the zooplankton, and the larger predatory hunters that feed on the smaller fishes. We have already met the mackerel and herrings of cold seas; warm seas too abound with epipelagic or surface-feeding fishes – fast-moving, sometimes highly-coloured, always restless and active in their unending search for food.

Liveliest and most elegant of the surface fishes are the many kinds of flying fish that occur in all the warm oceans of the world. There are about 120 species in all, and they feed mainly on small animals of the plankton and nekton.

When a shoal of flying fish is attacked, perhaps by a group of dolphins or other predators, the fish rise swiftly to the surface and shoot through it, spreading their large pectoral and pelvic fins as they clear the water. Often they skim along the surface, propelled by the lower lobe of the tail,

Left:
Radiolarians are tiny unicellular animals that swarm at the ocean surface. Strands of sticky material stream through the pores in their glassy shells, making a net to catch tiny algae and other prey. After death the shells sink to form an insoluble mud or ooze on the ocean floor.

Below left:
Great Frigate Bird of the Galapagos Islands. The male has inflated his crimson throat sac to entice the female to his nest for breeding. With a wing span of 2·5 metres (8 feet), Frigate Birds are splendid fliers. They catch Flying Fish on the wing, and bully other sea birds to give up food that they have caught.

Below
Flying fish are surface-feeding fishes with elongated fins. Chased by dolphins or other predators they leap from the surface and glide, sometimes 100 metres (330 feet) or more, with fins extended like wings. Changing direction in flight may allow them to escape from their predators.

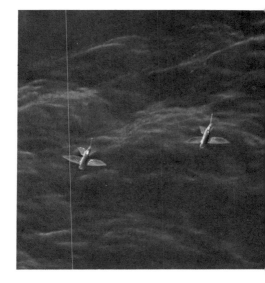

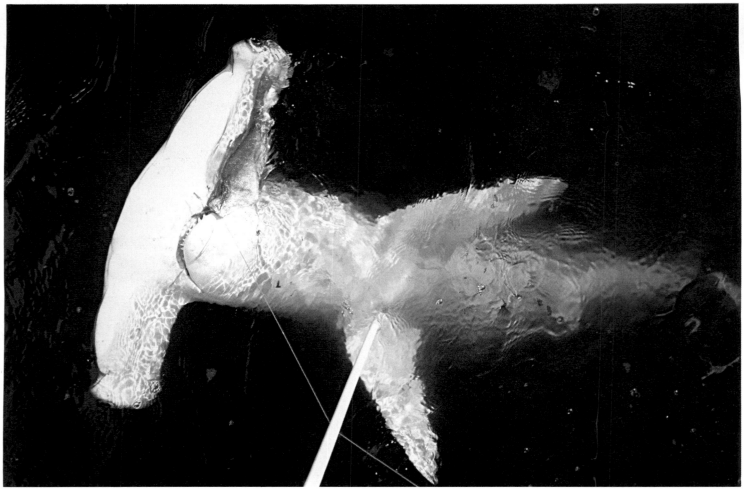

before rising clear and gliding. They may stay airborne for several seconds, banking slightly to change direction before diving in again. Their flights seem to be a way – and probably a very effective way – of avoiding predatory fish in the sea. However, while airborne they may fall prey to fast-flying frigate birds (*Fregata* species) – seabirds that snap them up with consummate skill as soon as the flying fish appear above the waves.

Fastest moving of the surface fishes are the tropical marlins (*Makaira* species) – long, slender fish, some of which exceed 3·5 metres (11 feet). The head is extended into a pointed 'marlin-spike' or spear, which seems to be used both for piercing large fish and for clubbing small ones. Almost as long and generally heavier in build are the tunas – Bluefins, Yellowfins, Bigeyes and Albacores, all of the family Scombridae. These are large, superbly streamlined fish with powerful muscles and scythe-like fins, widespread in temperate-to-warm oceans. The largest reach three metres (10 feet) in length and weigh over 150 kilograms (330 pounds); like marlins, they are reported to achieve speeds of 70 to 80 kilometres (40 to 50 miles) per hour.

Bluefin Tunas or Tunnyfish (*Thunnus thynnus*) of the Mediterranean and Atlantic range as far north as the Faroe Islands and Iceland in warm summers, and are often quite common off western Europe. It has been shown by marking them, that they migrate in shoals across the length and

breadth of the Atlantic from Florida to Norway. Albacores (*T. alalunga*) and Yellowfins (*T. albacares*) occur in both the Atlantic and the Pacific Oceans, and are equally wide-ranging. They probably follow the movements of the smaller shoaling fish which are their main diet. Tunas hunt in groups, swimming round the shoals to form dense concentrations and then charging through. Marlins tend to be solitary, though their hunting technique is similar.

Warm tropical surface waters are the favourite hunting grounds of the big sharks: the Great Whites (*Carcharadon carcharias*), Hammerheads (*Sphyrna lewini*), Blue Pointers (*Isurus glaucus*), Tiger Sharks (*Galeocerdo cuvieri*), Threshers (*Alopius vulpinus*) and many other species that have long been regarded as the villains of surface waters. The biggest reach 14 metres (45 feet) in length, but these are rare. Large sharks generally seldom exceed three to five metres (10 to 15 feet).

Predatory sharks of this size-range are found in all the oceans, mostly in the tropics and sub-tropics. Though mainly fish eaters, their preference for warm water often brings them close inshore, where they come into conflict with Man. Occasionally they attack bathers, though Man is not and never can be an important part of the diet of any shark. Their appetite is enormous. A Great White Shark caught in Port Jackson, Australia, which had been scavenging offal thrown into the sea, had in its stomach eight legs of mutton, 130 kilograms (280 pounds)

of horsemeat, half a pig, half a dog and a ham, together with sacking and a metal deck-scraper that some sailor had thrown overboard.

Big sharks usually hunt alone, or in loose bands that scatter over a wide area of sea. When food is found – perhaps a shoal of fish or a dying whale – the attacks of one shark attract others to the scene. Both the agitation and the scent of blood in the water are attractants, driving the sharks to a frenzy of cutting and tearing until the prey is destroyed.

Not all the surface-living sharks are hunters and killers. The biggest of all – Whale Sharks and Basking Sharks, some of which grow to 16 metres (50 feet), feed mainly on plankton and small fish, which they filter from the sea like enormous vacuum-cleaners. Their teeth are tiny and harmless. They are non-aggressive and completely inoffensive to larger creatures, and cruise lazily at the surface with only the tip of dorsal fin and tail showing. Widespread in all the oceans, they are quite common off western Europe, where they are sometimes hunted for their oil and their tough leathery skin.

Manta Rays (*Manta birostris*) – huge kite-shaped tropical fish six to eight metres (20 to 25 feet) across – are another form of large plankton-feeder, harmless to anything except the tiny creatures of the zooplankton that drift into their mouths as they flap and glide through the surface waters. They are related distantly to sharks, more closely to

the smaller skates and rays of the seabed. Mantas are black and sinister in appearance, but have done nothing to merit their alternative names of 'devil fish' and 'vampire fish'. They may even be playful, and sometimes can be seen leaping and somersaulting on the surface of the ocean like demented blankets, smacking hard on the water with their vast triangular fins. This

may be courtship or feeding behaviour. Though mantas are not uncommon in warm open seas, we know very little about their habits or behaviour.

Whales too can be playful and lively at the ocean surface. Though fish-like in shape, with fin-like flukes as forearms and a broad-bladed tail, whales are warm-blooded mammals with intelligence far above that of the fish with which they consort. There are about 70 species, divided into two great groups: the Mysticetes or whalebone whales, and the more varied and numerous Odontocetes or toothed whales. All whales spend most of their lives at the surface. As air-breathers, they cannot remain below indefinitely, though some can dive to depths of hundreds of metres and hold their breath for an hour or longer.

Mysticetes live very much at the surface. In the roof of their mouth they carry a mat of bristly whalebone (baleen) – like very tough hair or shredded horn – and through it they filter sea water to extract the zoo-plankton which is their main food. A big mysticete – a Blue Whale (*Balaenoptera musculus*) 30 metres (100 feet) long, or a Finback Whale (*B. physalus*), over 20 metres (65 feet) long, has an enormous mouth extending along a quarter of its length, and takes in sea water in hundreds of litres at a time. These whales feed mostly in cold waters where the plankton is richest, moving poleward in summer to catch the rich harvest of swarming crustaceans, and into warmer water in winter for mating and producing their calves. A new-born Blue Whale is eight metres (25 feet) or more long. Though thin, it may weigh over two tonnes. For the first few months it feeds entirely on its mother's milk, growing rapidly and fattening as it swims with the herd toward the summer feeding grounds.

Odontocete whales, generally smaller than mysticetes, are mostly less than 12 metres (40 feet) long. Sperm whale bulls (*Physeter catodon*) may reach twice *this* length, but are exceptional. Their cows seldom reach 10 metres (30 feet), and bull Killer Whales (*Orcinus orca*), the next smallest, average about the same length.

There are several species of beaked whales measuring between five and 12 metres (15 and 40 feet); but most of the dolphins and porpoises – which make up the bulk of the Odontocetes – are less than

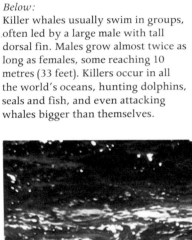

Below:
Killer whales usually swim in groups, often led by a large male with tall dorsal fin. Males grow almost twice as long as females, some reaching 10 metres (33 feet). Killers occur in all the world's oceans, hunting dolphins, seals and fish, and even attacking whales bigger than themselves.

Left:
Like a large sea-going tortoise with flippers and tough, leathery skin, the Leatherback Turtle spends practically all of its life in water; only the females come ashore for a few hours each year to lay their eggs.

Below:
Biggest of all the seals, the bull Elephant Seal may weigh over four tonnes. The 'trunk', an inflatable nose, is blown up when the animal is defending its harem of females on a breeding beach; it makes him look bigger and gives him a more fearsome roar.

five metres (15 feet) long. The smallest porpoises do not exceed two metres (six feet) when fully grown. Toothed whales are without baleen, and all have at least a few teeth in their jaws.

Fish-eating dolphins usually have most, sometimes as many as 80 sharp little teeth in either jaw. Killer whales and many porpoises have fewer, stronger teeth, suitable for tearing up carcasses of the large fish, seals and other big animals that are their prey. Sperm whales have teeth only in the lower jaw; beaked whales may have only a single pair of teeth in the lower jaw and none in the upper. Sperm whales and beaked whales seem to feed mainly on squid – soft-bodied, slippery creatures that live well below the surface. How these strange dental patterns help in catching and eating squid is not known.

Whales are the commonest and most widespread mammals of the open ocean. Seals, too, can be plentiful, especially in offshore waters where upwelling brings concentrations of plankton and fish. Including the walrus and the several kinds of fur seals and sea lions, there are over 30 species in all, ranging in size from the great Southern Elephant Seal (*Mirounga leonina*) – some bulls of which are over six metres (20 feet) long – to small fur seals a little over a metre (three feet) in length.

Seals, like whales, are warm-blooded and air-breathing mammals. Unlike whales, they spend part of their lives on land, coming ashore or on to sea ice to rear their pups. Most are fish-eaters. A few, like the Crabeater Seal (*Lobodon carcinophagus*) of the Antarctic pack ice and Ringed Seals (*Pusa hispida*), include large crustaceans of the zooplankton in their diet. Walruses (*Odobenus rosmarus*) and Bearded Seals (*Erignathus barbatus*) of the Arctic feed on clams and other creatures dug up from the seabed. Their sensitive whiskers help them to detect prey in waters too muddy to see through. Many seals make long seasonal

migrations in surface waters, spending several weeks or months each year out of sight of land.

Not unlike seals in their need to breed ashore are the five species of turtles that live in the open sea. Four species – the Green (*Chelonia mydas*), Hawksbill (*Eretmochelys imbricata*), Loggerhead (*Caretta caretta*) and Ridley (*Lepidochelys olivacea*) have hard shells like tortoises; the fifth – the Leatherback (*Dermochelys coriacea*) – has a leathery skin. Marine turtles are green or dull brown, with long, elegant flippers and a beak-like snout. Green turtles feed mainly on sea grasses that grow in shallow water; the remaining species are

Top:
Storm petrels flutter lightly over the oceans, picking plankton and shrimps off the surface without wetting their plumage.

Above:
Tropic birds are oceanic birds of warm waters that plunge to snatch fish and squid from the sea surface. The long tail feathers wave like streamers in flight. The feet, small and set well back, are almost useless for walking or standing; tropic birds nest mainly on cliffs.

mostly carnivorous, feeding on jellyfish, siphonophores and fish. The females of all five species drag themselves ashore to lay their eggs in deep holes which they dig above the high-tide mark on sandy beaches.

Turtles are some of the very few reptiles that go to sea. Others include large crocodiles of Australia and South-east Asia, that often appear in estuaries and mangrove swamps, and sea snakes, most of which live in the Indian and warm Pacific Oceans.

Seabirds
The biological wealth of any patch of ocean can usually be judged best by the number of seabirds feeding from it. Highly mobile,

resourceful and opportunist, birds use a great deal of energy in their daily lives, and can never afford to stray far from a source of food. They watch each other closely; where one feeds, others gather round and feed too.

Some of the biggest assemblies of birds occur at sea. Flocks numbering millions of individuals, often of only a very few species, may accumulate over the rich feeding grounds in Antarctic waters, and in tropical seas – for example the coast of Peru – where upwelling brings nutrients to the surface.

Seabirds feed in many different ways. We have already met frigate birds – long-winged black oceanic birds that swoop low

over tropical seas and catch flying fish (and occasionally flying squid) on the wing. Flying fish and others of their size can also be caught by boobies (*Sula* species) and gannets (*Morus* species) and tropic birds (*Phaethon* species), that plunge into the water after them, often from heights of several metres. Shearwaters (*Puffinus* species) settle on the water and then dive, swimming actively after their prey with feet pedalling and wings part-extended to form hydrofoils.

Cormorants and shags (*Phalacrocorax* species) dive deep, again using their big webbed feet as underwater propellers and taking fish from depths of 20 or more metres (65 or more feet).

Penguins have given up flying altogether – and almost lost their wings – to specialize in swimming and diving. Some feed mainly on plankton at the surface, but the largest species, Kings (*Aptenodytes patagonica*) and Emperors (*A. forsteri*) dive to depths of several hundred metres after large fish and squid.

The smaller seabirds are entirely surface feeders. Storm Petrels dance over the waves, barely touching them with their tiny webbed feet, but picking plankton constantly off the water. Diving petrels (*Pelecanoides* species) dive through the waves with wings whirring, emerging on the other side with a small fish or particle of zooplankton. Prions (*Pachyptila* species) settle and sift the sea water through their bills, holding back particles of food. Terns hover and dip; closely-related skimmers or scissor-bills (*Rynchops* species) fly over the water with lower bill trailing, picking up crabs and small fish as they go.

Big seabirds usually settle when they eat. Albatrosses swoop back and forth over hundreds of kilometres of ocean, then settle when they find food close to the surface and dip with their long neck and bill. So where several species of seabirds are feeding together, they are likely to be catching and eating different foods; theirs is a rich table, and their various methods of hunting ensure that many can feed from it at once.

Above:
Penguins are flightless seabirds; with wings modified as paddles they swim and dive very efficiently, catching fish, squid and plankton. Insulated against the cold sea by fat and dense plumage, they are well equipped for living on cold lands.

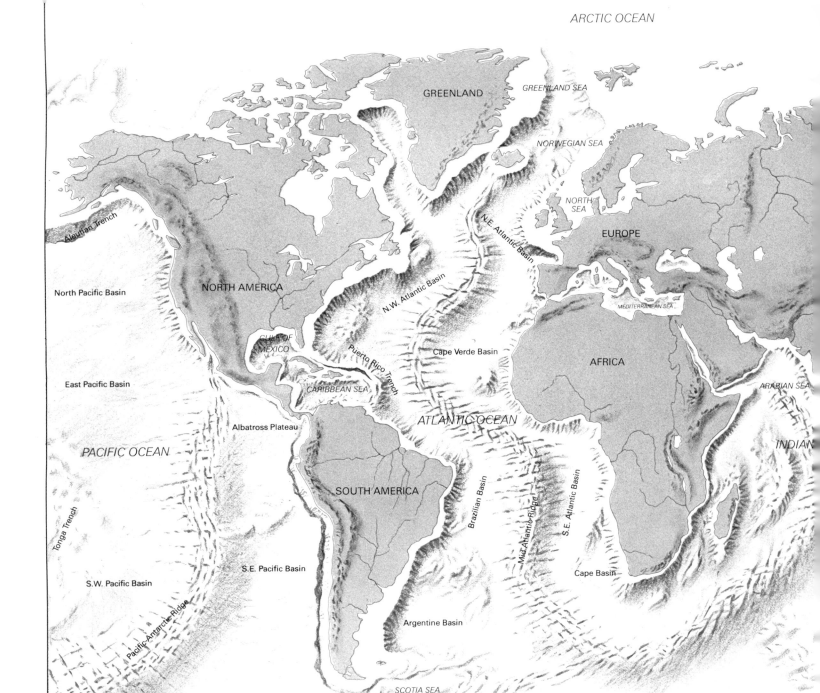

Labels on map:
ARCTIC OCEAN

GREENLAND
GREENLAND SEA
NORWEGIAN SEA
NORTH SEA
EUROPE
N.E. Atlantic Basin

Aleutian Trench
NORTH AMERICA
North Pacific Basin
N.W. Atlantic Basin
MEDITERRANEAN SEA

East Pacific Basin
GULF OF MEXICO
Puerto Rico Trench
Cape Verde Basin
AFRICA
ARABIAN SEA

CARIBBEAN SEA
Albatross Plateau
ATLANTIC OCEAN
INDIAN

PACIFIC OCEAN
SOUTH AMERICA
Brazilian Basin
Mid Atlantic Ridge
S.E. Atlantic Basin

Tonga Trench
S.E. Pacific Basin
Cape Basin

S.W. Pacific Basin
Argentine Basin
Cape Basin

Pacific-Antarctic Ridge

SCOTIA SEA

SOUTHERN OCEAN

Deep seas

Above:
The bed of the ocean. Once thought of as a lifeless muddy plain, the sea floor is now known to include mountain ranges and deep trenches. Living animals have been brought up from even the greatest depths.

Below the teeming, sunlit world of the ocean surface lies a darker, colder zone, far more difficult to reach and examine. Long after the major continents had been outlined by explorers, the ocean bed remained a mystery. About 150 years ago, biologists thought of it as a desert. Lack of light, they argued, and the enormous pressure of water would make all life impossible below a depth of about 550 metres (1,800 feet) – anywhere, in fact, beyond the edges of the continental shelves.

The *Challenger* expedition
The marine explorations of the mid-nineteenth century began to show life on the bottom of deep Norwegian fiords – even

sedentary animals fixed firmly to the sheathing of submarine cables, brought up for repair from depths of almost 2,000 metres (6,500 feet). It was partly to solve the problem of life in deep water that the *Challenger* expedition of the Royal Navy set out in 1872, and explored the oceans of the world for almost three years. By the end of the voyage, its scientists had shown that life existed in every corner of the ocean, down to the greatest depths (4,400 metres, 15,000 feet) that their clumsy equipment allowed them to penetrate.

For a long time after *Challenger* our knowledge of deep water life was restricted to fragments, mostly of dead and dying animals, brought up to the surface by

dredges and grabs. To the damage inflicted by the collecting gear could be added the damage due to changing pressure. For every 10 metres (33 feet) of depth, hydrostatic pressure increases by one atmosphere (roughly one kilogram per square centimetre or 14·72 pounds per square inch), and few fish survive being hauled up even from depths of a hundred metres.

Yet by the early days of this century, fish were found living at depths of 6,000 metres (20,000 feet) and more, and new forms of life were discovered every year from dredging in one part of the ocean or another.

The old methods of catching still help us to discover new animals and communities on the seabed, but today we have underwater photography and television too, allowing us to scan the sea floor and actually see what the animals are doing. We can even go down to the deep ocean floor in submersible craft to see for ourselves.

From all the evidence pieced together we can now be sure that life exists at every level of oceanic water, and all over the seabed. Living creatures have even been found in the Philippine Trench, a ravine in the floor of the western Pacific Ocean 11,500 metres (37,000 feet) deep. This is the deepest corner of the ocean so far discovered. You could drop in Mount Everest and stack seven Empire State Buildings on top, and still need a long ladder to reach the surface.

Views about life in deep water have changed, and so has our picture of the seabed itself. During the early days of marine exploration, soundings were taken with long lead-lines of rope or wire, and one deep-water sounding per day was a good rate of progress. Now oceanographers use sonar and other continuous techniques that draw profiles of the ocean bed, with far more detail than the sounding line could provide.

The seabed is much rougher and more varied than was thought originally, with enormous mountain ranges, chasms, steep-sided valleys and rolling hills. There are vast canyons, deeper and longer than the Grand Canyon of Colorado, off the mouths of some of the great rivers, and many tall mountains that sweep up in isolation from the plains around them. Some of the mountains reach up to the sea surface and form islands: Hawaii and the Azores Islands are really groups of underwater mountains – chains of volcanoes, some still active, that reach the surface and tower above it. Many

Top:
Scientists of HMS *Challenger*, a Royal Navy expedition ship, explored the oceans, pioneering methods of deep-sea sounding and dredging. Their results, filling fifty massive volumes of scientific reports, are the foundation of modern oceanography.

Above:
Surtsey, a volcanic island that grew out of the sea off the southern coast of Iceland between 1963 and 1967. It stands toward the northern end of the Mid-Atlantic Ridge, a line of fracture in the crust of the earth. Already plants, insects and seabirds have begun to colonize it.

Right:
How volcanic islands are formed.

Below:
The distribution of animals in deep
water.

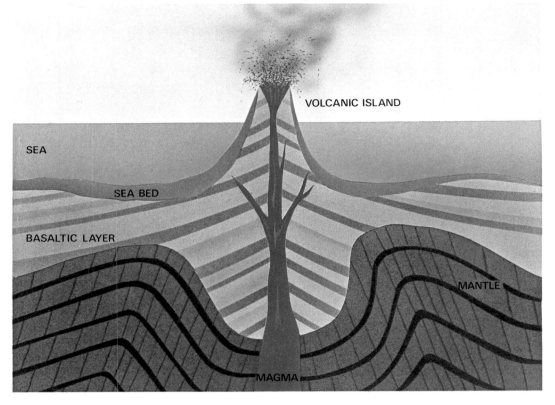

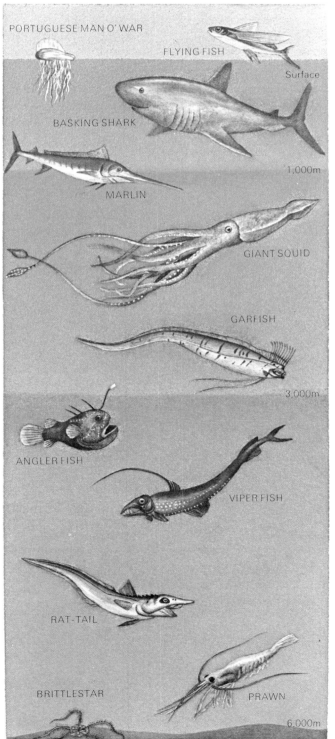

sea mountains are not quite tall enough to
make islands. One in the Pacific Ocean rises
eight and a half kilometres (five miles) from
the seabed – almost as high as Mount
Everest – but still does not reach the ocean
surface.

The ocean zones

Biologists divide the ocean into convenient
horizontal zones. The world of the plankton, called the *epipelagic* zone, extends
down to 250 metres (800 feet) or so. Below it,
from 250 to 1,000 metres (800 to 3,000 feet),
lies the *mesopelagic* zone, and below that lies
the *bathypelagic* zone down to a depth of
4,000 metres (13,000 feet). Deeper still, in
discontinuous patches separated by the
underwater mountain ranges, lie the basins
of the *abyssal* zone, extending down to the
greatest depths of the ocean.

Taking a bathyscaphe (a submersible
chamber) and descending steadily through
these zones of life, we pass quickly beyond
the cloudy, dappled plankton into a twi-
light blue-green zone. Immediately we see a
change of colour in the animals around.
Where epipelagic animals were striped and
barred, in every colour of the rainbow,
mesopelagic animals tend to be red or
orange. Fish are black or silvery, sometimes
reflecting light from the bathyscaphe like
brilliant mirrors. In the dim, natural light
outside, both orange and black pigments
make their owners practically invisible,
and even the silvery mirrors are lost when
the light is diffuse. There is still a faint
glimmer of sunlight even at 600 metres
(2,000 feet), but it seems to be coming from
all sides. As we descend further we become
aware of a new kind of light: practically
every creature we see carries tiny phos-
phorescent lights of its own, which shine

through the darkness when the last traces of daylight have disappeared.

Light organs, or photophores, have many different functions and arrangements. Some shine all the time; others can be dimmed by curtains of black pigment, or snuffed out by controlling the blood supply. Though at first glance they may seem dangerous because they give away the position of the bearer, they obviously confer advantages as well.

Rows of photophores on the lower surface of an animal disrupt its outline when it is seen from below. A faintly-lit organism shows up less than an all-black one against the faint light from the sky. *Vinciguerria*, and many other mesopelagic fish, are equipped in this way. Like other predators of this zone, they migrate up into the plankton each evening and descend at first light, so are often at risk from hungry jaws below them. Other creatures use photophores in patches to break their outline, making it difficult for a hunter to know where to bite.

Others again carry them only on extensions of their antennae and tail, giving themselves a completely spurious outline that predators must find disconcerting in the darkness.

Photophores are often used as lures. Several kinds of angler fish wave them on stalks in front of their mouths. Other species (*Astronesthes*, for example) have illuminated chin barbels, while others like the viper fish (*Chauliodus*) carry lures on their dorsal fin and inside the mouth. Flashing lights may be used to attract males and females of the same species to each other in the mesopelagic gloom; and brilliant, timely flashes may help prey to escape as the jaws are closing about them. That photophores help predators to find prey must also be true.

Many fish of this zone have large, owl-like eyes, often upward pointing, with retinas far more sensitive than our own to pinpoints of light against a dark background. Some even carry 'headlamps' —

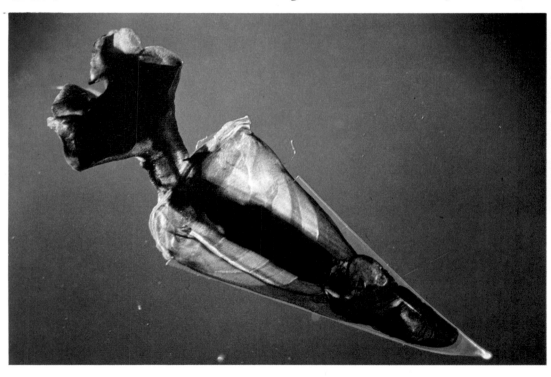

Above:
A pteropod or Sea Butterfly (*Diacria trispinosa*) of the Atlantic Ocean. Pteropods are related to snails; of the two major groups, one even retains a shell. Wing-like extensions of the foot flap to keep them at the right depth.

Above left:
A pteropod of mid-to-deep water (*Clio* species). Like *Diacria*, these belong to the Thecosomata or shelled group of pteropods. Seldom more than 2 to 3 cm (one inch) long, they feed on microscopic particles sifted from the water around them.

Left:
Snipe eel (*Nemichthyes* species), a deep-water predatory fish. The large eyes and elongated jaws characterize the family, but little is known of their way of life.

Above:
A deep water squid (*Gonatus* species).
Moving by jet propulsion, squid are
found at all depths of the oceans.
Some of the deep-water species carry
lights and the largest ones grow to 15
metres (50 feet) or more in overall
length.

Opposite, top:
Like their counterparts in shallow
water (page 52), deep-sea angler fish
dangle an attractive lure above their
mouth. In the pitch-dark depths of
the ocean several species carry
brilliant lights.

Opposite, bottom:
A snapping shrimp (*Alpheus* species)
of deep water. The large claws
include a trigger mechanism that
clicks loudly when discharged,
stunning nearby fish and
immobilizing them.

arrangements of photophores near the eyes,
with reflectors behind them. These are
probably used to locate prey just a few
centimetres in front of the jaws.

Mesopelagic fish are often plentiful; suf-
ficiently so for their swim-bladders collectively to affect the echo-sounders of ships
like an almost-solid surface. This spurious
echo, recognized long before its cause was
known, was called the 'deep-scattering
layer'. It puzzled scientists by moving up
each evening and down in the early morn-
ing, though trawl nets put through the
scattered fish usually come up empty.
These fish are preyed on by each other, and
by large, lively hunters such as sharks that
catch them during their nightly excursions
into the plankton.

The mesopelagic zone is also the home of
cephalopods, including the giant squid that
for centuries have formed one of the many
bases for sea monster legends. Squid are
simply high-speed molluscs, related to
Octopus and *Sepia* of shallow waters.
Several species in a wide range of sizes
occur in deep water, living close to the
bottom and probably also in mid-water.
Other eight-armed and ten-armed octopus-
like creatures live with them, all lively
predators on the fish and bottom-living
creatures that are found in this broad band
of ocean.

Most attractive of the free-swimming
molluscs are the Pearly Nautilus, sole sur-
viving species of the once dominant

ammonites whose coiled shells so often
occur in Mesozoic rocks. *Nautilus pompilius*
still retains a small shell, which in life is
partly hidden by a mantle. It serves as a
buoyancy chamber, keeping the animal at
depths within the mesopelagic zone. A
nautilus swims by jet-propulsion, forcing
water through a muscular funnel in its
mantle. It feeds by holding fish and other
prey with its many tentacles, and biting
them with a sharp, parrot-like beak. The
zebra-striped shells, up to 30 centimetres
(12 inches) in diameter, are prized by collectors, though the nautilus live too deep to be
hunted. Only dead or dying specimens are
usually caught.

Squid are by no means uncommon in the
oceans. The larger species are eaten by
several kinds of toothed whales, the smaller
ones by dolphins, many species of fish,
seals, birds, turtles and a host of other
predators. Only Man seems to have diffi-
culty catching them. Most of the squid
examined by biologists have been re-
covered from the stomachs of other animals.
They are probably too widely dispersed
and too alert to be caught in slow-moving
mid-water trawl nets, or in coarse nets
streamed through the plankton at night.

Many forms of squid live in mid-water,
including *Lycoteuthis diadema*, a tiny
species decorated with photophores, and
Calliteuthis reversa, also small and illuminated, with eyes of unequal size. *Onycho-
teuthis banksi*, a slightly larger species, lives

in deep water during the day but feeds in the plankton at night. Like other small squid, it lives in shoals, and sometimes makes its presence known by flying through the air. As with flying fish, this is thought to be the way they avoid fish predators.

Among the larger squid are several species of *Architeuthis*, which are occasionally washed ashore dead or found, usually in pieces, in the stomachs of Sperm Whales. The largest have bodies measuring six to seven metres (20 to 23 feet) and even longer arms. Smaller ones often leave the marks of their hooked suckers on the skin of Sperm Whales. Squid are believed to feed mainly on fish, stalking rapidly with a form of jet-propulsion and grasping their prey with their long, many-suckered tentacles. Most of the mid-water forms have well-developed eyes, and batteries of photophores on the body and tentacles.

In the deeper waters of the mesopelagic zone and through the bathypelagic zone, the population of fishes thins, and both pelagic and bottom-living species become more bizarre and outlandish in appearance. If we lower a camera or bathyscaphe we see very few at first. They are probably widely scattered in their constant search for food. But lowering a bait as well soon attracts a crowd. Whatever the ocean, it is a remarkably similar crowd that gathers. Large numbers of middle and deep-water fish belong to one or other of the families Macruridae (the rat-tails) and Brotulidae – slender creatures with large heads and thin, tapering tails.

There are also ray-fins (Sudidae), with long stilt-like fins that support them on the soft mud, viper fish and hatchet fish (Stomatoidae) with huge jaws and vastly extensible stomachs, angler fish (Ceratioidae) with their dangling lures, and

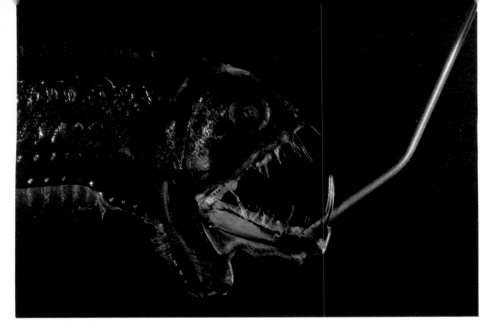

gulper eels (Saccopharyngidae) which seem to be nothing but a pair of huge jaws with an eel-like tail behind.

Whether bottom-living or pelagic, these curious fish have many features in common. They are mostly small – less than 30 centimetres (12 inches) long – and brown or black with silver mirroring or a sprinkling of photophores. Often they are rather sluggish, with small muscles. Their eyes are usually tiny, and the bodies of many are extended into whip-like tails. These sometimes bear prominent lateral line organs – the rows of sensory pits which fish use to detect movement of prey at a distance. They are usually thin, sometimes with hardly any abdomen between head and tail. Nearly all have disproportionately large jaws, that unfold, like pen-knives, to engulf creatures broader and deeper than themselves. Some make noises with their swim-bladders or jaws, perhaps to attract others of their own species.

Their way of life seems to be to move slowly in search of food, which can only be rare in the deep waters of the ocean. Then they grab and engulf it. Whether it be a carcass from above, or another fish in an unguarded moment, they take anything that will fit into their extraordinarily extensible jaws. Incurving teeth in serrated rows ensure that food, once grasped, can travel only one way. So a deep-water fish 12 centimetres (five inches) long can catch and ultimately swallow another almost twice its length. Then, presumably, they live quietly waiting until the next morsel of food comes along, or their own unguarded moment gives their neighbour his next meal.

Little is known about the breeding of deep-water fishes, but one group, the angler fish, has solved the problem of finding partners in an interesting and unusual way. Biologists were puzzled to find that all the large, mature angler fish they caught in deep water were females; males seemed to have disappeared. Then it was discovered that most of the females had one or more tiny males – easily mistaken for flaps or threads of leathery skin – attached to them. Males apparently find their partners and clamp on tightly, the tissues of their mouth fusing with the female's skin. Her blood vessels supply food, and the male produces sperm which fertilizes her eggs as they are released into the water. Other species change sex, starting as males but turning into females later in life; others again can function as male and female simultaneously, and may even be self-fertile.

Material of the ocean bed

The bed of the ocean is littered thickly with sediments of various kinds, providing a substrate for a host of bottom-living animals. Under the warm oceans, down to depths of about 4,000 metres (13,000 feet), the sediments are mostly calcareous remains of foraminifera – microscopic single-celled animals of the plankton. Under polar oceans, siliceous shells of diatoms (microscopic plants) predominate. The deep-ocean bed surrounding Antarctica is littered many metres deep with white diatomaceous ooze. In very deep water, only reddish or brown clays carpet the ocean floor, littered with sharks' teeth and other completely insoluble organic remains.

Close to the mid-oceanic ridges, where the sea floor is relatively new, the sediments lie thin; bare pillow lavas and other volcanic rocks may form a hard surface. In older parts of the ocean, organic sediments alone may reach thicknesses of several hundreds of metres. Near the continental shelves and other submarine slopes, sediments from the land may accumulate through slumping; silts and muds pour in avalanches on to the ocean floor, gouging great canyons and spreading huge, flat deltas of sand and mud. These, then, are the various kinds of substrate available for the bottom-living animals of the oceans.

There seems to be life all over the ocean floor, generally richest in the shallows and poorest at great depths. In depths of about 1,000 metres (3,300 feet) the rain of organic sediments from the plankton is plentiful enough to spread a fairly rich carpet of debris over the bottom. Filter-feeding animals – sponges (especially the delicate siliceous glass sponges), bivalved molluscs, sea pens, sea fans, sea squirts, and many other creatures familiar from shallower water – may cover the seabed, packed together, jockeying for position with starfish, brittlestars, echinoids, polychaetes, snails and sea cucumbers that feed on or just below the surface of the mud.

In deeper water, the rain from above is poorer in both quantity and quality. Large particles are snapped up before they can sink to the bottom. Smaller particles lose all their nutrients to bacteria or to the surrounding water, and are little more than husks when they reach the seabed. There is

a lesser rain of faeces and dead material, however, which is enough to support lesser communities of mud-ingesting sea cucumbers, fishes and snails, and scattered brittle-stars and sea anemones. Fishes and squid are still quite common at depths of 4,000 to 5,000 metres (13,000 to 16,000 feet). A brotulid has been caught at over 7,000 metres (23,000 feet) and ray-fins have been seen at similar depths, and a flat fish, probably of the sole family (Soleidae), presented itself outside a bathyscaphe at 10,000 metres (32,000 feet) in the bottom of a submarine trench off the Marianas Islands. From the similar trench off the Philippine Islands a research ship hauled a catch of 25 sea anemones, over 70 sea cucumbers, and a selection of bivalved molluscs, worms and amphipods – representing the deepest-living communities

Man and the oceans

For as long as he has known the sea, Man has made full use of its riches. In the hunting and gathering stages of human evolution, the inter-tidal zone provided rich pickings for nomadic bands. Some of the oldest human settlements we know are marked, not by pottery, axes or the remains of dwellings, but by piles of shells and

bones – the shells of oysters, mussels and other inter-tidal molluscs, and the bones of fish caught close to the shore. As far apart as Britain and Australia, Japan and Cape Horn, primitive man discovered the bounty of the seashore and shallows, and lived well on it throughout the year.

Rich pickings

We take an enormous range of foods from the sea. For hundreds of generations we have dived to the seabed for clams and mussels, set pots and traps for crabs and lobsters, and built weirs across estuaries and small bays to take char, salmon and 'whitebait' (larval fish of several species) at the start of their journey from the sea to fresh water.

We have dug for ragworms (still regarded as a delicacy in some parts of the world), made ponds to catch shrimps, and worked through beds of cockles and oysters to find the largest and most succulent. In the days before farming and the domestication of animals, these must have provided important sources of protein for coastal dwellers, whose small, roving populations would have made very little impact on the natural communities. Seafoods that we now prize as delicacies must once have sustained human bands against starvation. Even today there are whole human populations – Polynesian islanders, for example – who rely heavily on the sea for their daily supplies of protein, and seldom see any other kind of meat.

Settlement, farming and stock-rearing brought other sources of protein for advanced societies, but fish – first from shallow waters and then from the deep – has always been an important component of civilized human diet. Sea fishing was one of our earliest commercial skills. Egyptian fishermen 5,000 years ago knew the use of the seine or drag-net, a long net that could be operated from land, or from reed-boats and coracles. This was the fishing method of Biblical times. It was probably the Phoenicians who first brought the technique to Europe over 1,000 years ago, and started the shallow-water fisheries that proved so important in European welfare and development.

Herrings were one of the first fishes to be exploited in the North Sea. There are records of salted fish being exported in bulk from Scotland to the Netherlands as early as the ninth century, and Great Yarmouth and Lowestoft were well established as herring ports by the eleventh century. By the twelfth and thirteenth centuries fishing

was well established in offshore waters; the North Sea was being exploited by both net and line fishermen, who served growing local communities in Britain and maritime Europe. By the fifteenth and sixteenth centuries, commercial fishing for cod and haddock extended out into deep waters.

Hand-line fishermen in Henry VIII's time (early sixteenth century) were sailing from Britain as far afield as Iceland. Beam-trawling began in the North Sea perhaps as early as the fourteenth century, but was unpopular among fishermen because of the damage that might be done to the fishing grounds.

By the seventeenth century, as the demand for fish increased, trawling was more widely used, though still regarded with suspicion by fishermen and authorities alike.

Meanwhile, similar techniques were being developed for fisheries right across the world. Many techniques for catching fish seem to have occurred spontaneously to widely-scattered populations of fisherfolk,

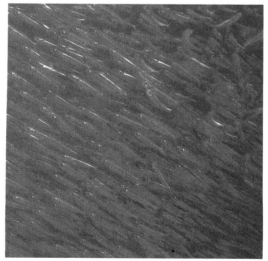

though seamen during the great voyages of exploration in the fifteenth to eighteenth centuries were probably responsible for the spread of several. Arabs carried their methods of fishing across the Indian Ocean. Maoris and other Polynesians spread their methods throughout the southern Pacific, while Japanese and Koreans traded ideas in

Opposite, top
Once a popular and cheap food – now still popular but expensive – the Common Oyster (*Ostrea edulis*) growing at low water springs on the mud of an English estuary. Oysters have long been farmed, pointing the way for further research and development in cultivating stocks of sea foods.
Opposite, bottom:
Fish that are plentiful but seasonal have to be preserved for year-round supplies. Here salmon are drying in the sun – an age-old method of preservation still in use today.
Above:
Techniques of netting fish from river banks, shores and small boats are well over 5,000 years old. Shown here are Egyptian and Byzantine fishing boats.
Left:
Shoals of fishes like these Herrings (*Clupea harengus*) have for many millions of years been hunted by larger fish, whales, seals and seabirds. Man is a relatively recent predator, but has already exterminated huge populations overfishing.

the north.

In nineteenth century Europe, where the industrial revolution was spreading prosperity and populations were increasing rapidly, the pace of commercial fishing accelerated. New roads, and the new-fangled railways, created new inland markets for fish, which retained its value as a cheap form of protein. Catches increased enormously. Trawling, now regarded as a necessity, spread from the Thames estuary to shallow areas all over the North Sea. With the advent of steam power, bigger and better trawls were devised, larger and faster ships were designed, ice came into general use as a preservative, and British and European trawlers spread far and wide across the North Atlantic, into the Arctic Ocean.

Then, during the early days of the present century, it became clear for the first time that local stocks of many kinds of food fishes were declining, quite simply through overfishing. Both quality and quantity were falling. The spread of fishing to more distant grounds took some of the pressure off local stocks and coped for a short time with the ever-increasing demand for fish.

But then the best far-distant fishing grounds too began to show signs of over-exploitation, bringing keener competition for the few unexploited stocks remaining. Once thought of as an infinite resource ('There are as many good fish in the sea as ever came out of it,' said our grandfathers) sea fisheries were found to be finite, and in need of international protection.

Protection of sea life

However, it has proved very difficult to devise laws for the protection of fish stocks. Except for those living close inshore, it is difficult to say who owns fish, and who has the right to protect them. The nations cannot agree on how far out into deep water protective measures should extend, and the

fish of deep water, well away from land, are still there for anyone and everyone to take. Now fleets of super-catchers, with factory ships in attendance, are exploiting deep-sea fishes all over the world, often processing them on the spot into meal for animal feed and fertilizers. And we can be sure that the catches will be controlled, not by the level of exploitation that these stocks of fish can stand, but by the increasing demands throughout the world for the products of the fishery.

Seals have been a target for human predators over many centuries. Those that came into contact with primitive societies in northern oceans were killed for their meat, blubber and skins, though their stocks as a whole were seldom reduced by the small numbers taken. A demand for furs – used for clothing and decoration in more sophisticated parts of the world – put additional pressure on northern stocks of seals, and spread to the far south. Captain Cook, during his southern voyages of exploration in the late eighteenth century, reported the presence of many fur seals on islands in the southern oceans. British and American sealers that visited those islands during the next half century took hundreds of thou-

Below:
South African Fur Seal and her pup (*Arctocephalus pusillus*) at their northern limit of breeding – Cape Cross, South West Africa. Fur seals have been exploited for their skins for many centuries, often with complete destruction of stocks. Sensible management is a possible alternative.

sands of fur seals – bulls, cows and pups indiscriminately – wiping out populations completely and then moving on to discover new islands and new populations, and destroying those as well. Seals are still at risk, and will remain so for as long as we think it smart to wear sealskins ourselves.

Whales too have been exploited over the centuries in similar destructive ways. Whaling, like fishing, began in shallow water. Dolphins and other small-toothed whales were driven ashore or netted, and slow-moving Right Whales, Grey Whales and Humpbacked Whales were harpooned and beached. The oil, meat, sinews and bones from their carcasses must have helped many small coastal communities to survive during long, hard winters. By the sixteenth century, whalers had taken to sailing ships and were travelling long distances to harpoon whales. They were especially interested in whalebone – which was used for making all kinds of household goods from brooms to corset-stays – and oil, used as fuel, lubrication, and a dressing for leather.

Up to 1875 the slow-moving Right Whales, Humpbacks and Sperm Whales were the main prey of the whalers. These could be harpooned from open boats, and stayed afloat long enough after death to be towed back to the mother-ship and stripped down.

Then, with the invention of the explosive harpoon, faster-moving whales could be tackled. Faster ships were employed to catch them, and both floating and shore-based factories were set up all over the world to process the whales as quickly as they could be brought in. In the early twentieth century, factories were established for the first time in Antarctic waters, and a whole new era of whaling began in the cold, rich ocean surrounding the southern continent. Now stocks of whales all over the world have declined, some almost to vanishing point. Again, it has proved almost impossible to devise laws covering a resource that nobody really owns, but everyone wants to exploit as quickly as possible.

Both fish stocks and whale stocks increase rapidly if they are left alone. This was shown very clearly during the two World Wars, when the main fishing and whaling fleets were put out of business. Stocks of their prey were much improved on the return of peace, yielding bumper harvests for several years. Many scientists

Above and left:
Southern Right Whales (*Eubalaena australis*) feed in the open sea but come close inshore, often into shallow bays, to produce their young. In the early days of whaling this made them particularly easy to kill. Now remaining stocks are protected by international agreement.

Below:
Seabirds suffer particularly from oil slicks. The oil sticks to their feathers so they can neither swim nor fly, and slowly poisons them. This Western Grebe (*Aechmophorous occidentalis*), caught by the 1974 oil disaster in San Francisco Bay, was treated successfully. Many thousands die each year.

and international law-makers see this as a reason for taking pressure off the stocks of food fishes, and suspending whaling altogether for at least ten years. This would give the hunted stocks a further chance to recover, and perhaps give us all time to make better laws for the sensible management of whatever remains.

But meanwhile the competitive hunt for fish and whales continues. Now a new resource – krill from the southern oceans – is just starting to be exploited by the fishing fleets of several countries. Rich in oils and proteins, this was the main food of the southern baleen whales. Since they have mostly been destroyed, there is plenty of krill for the taking. However, seals, fish and seabirds by the million take it too. Over-exploitation of krill by Man would be at the expense of the remaining wildlife of the southern oceans, and might well destroy the last possibility of whales being able to return to their former numbers.

To the other benefits the oceans have given us, we can now add oil, natural gas and minerals from the seabed. Like the rest of the benefits, we take them as fast as they are discovered. In return, we pour into the sea each year several millions of tonnes of rubbish – sewage, factory wastes, garbage and oil. We once thought its resources unlimited; now we act as though we thought its capacity for absorbing wastes is unlimited, but we may be wrong.

Though the sea has enormous volume, and a tremendous capacity for neutralizing, recycling and otherwise disposing of many kinds of organic and inorganic matter, it cannot be all-forgiving. Already the effects of our pollutants are visible in the open oceans far from land – globules of fuel oil in the mid-Atlantic, radioactive wastes and insecticide residues in the Antarctic oceans are clear examples, and the sea is absorbing much of the carbon dioxide and other soluble wastes that pour from our factory chimneys.

Rather than testing to destruction the sea's capacity and generosity, we would do better to find safer, more efficient and economical ways of disposing of wastes. Should we manage in our bumbling, greedy way to destroy the oceans, with their enormous capacity for living, we would certainly have succeeded in destroying ourselves as well.

Index

Textual reference is in Roman and captions to illustrations in Italic.